umbers, points, line**s, planes, space,** trian-
l**es, squares, spirals**, goo-
ols, primes, imagi**D0485581** rcles,
actions, transcend nega-
ve, golden rectangle, π, wavelets, numer-
ls, domains# **numbers** nities, fractals,
ermutations, operations, i, Pythagorean
heorem, hyperbola**&** finite, zero, sets,
roblems, irrationals, complex numbers,

other math ideas
roblems, Goldbach's conjecture, cycloid,
on-algebr **come alive** roots, radi-
als, parabolas, tetrahedron, geometric sol-
ds, limits, e, non-euclidean geometries,
uclid's 5th postulate, hypercube, absolute
alue, reciprocal, base, binomial, expo-
ents, square root, convergent series, geo-
netric mean, factor, catenary curve, geo-
netric series, arithmetic mean, function,
inear equation, quadratic equation, se-
uence, root, polar coordinates, radicals, si-
nultaneous equations, conic sections, mid-
oint, median, icosahedron, Feigenbaum
onstant, hexadecimal number, variable,

numbers
and other
math ideas
come alive

theoni pappas

WIDE WORLD PUBLISHING

Portions of some stories in this book have appeared in previously
published works by Pappas.

Wide World Publishing/Tetra
P.O. Box 476
San Carlos, CA 94070
web sites:
http://www.wideworldpublishing.com
http://www.mathproductsplus.com

Printed in the United States of America.

First Printing March, 2012   —   ISBN: 978-1-884550-63-8

First e-book format edition 2012 — ISBN: 978-1-884550-64-5

To
my parents
Frances and James
and
Elvira
for her encouragement
with these stories

— Contents —

Introduction

Contrary to popular belief, math is not just a bunch of math facts and concepts that never change or evolve. Every math idea or problem takes time to evolve while it percolates in the mind of the mathematician. Imagination and mathematics have always been closely connected. In fact, mathematics is pure fantasy but with one very important element...logic. Each new idea that is created or uncovered must follow logically from previously established concepts, regardless if they are some new type of number, a definition, a theorem, or other math notion. They must be shown to be consistent with previous ideas. Looking at the history of mathematics we realize that new ideas were oftentimes not readily accepted. Some were even considered blasphemous, ridiculous or downright dumb. Look at Leopold Kronecker's vicious and insidious attack on George Cantor's innovative work on set theory and transfinite numbers, consider mathematicians negative reaction to the negative integers when they were first introduced to the European continent or how the Pythagoreans tried to keep the discovery of the $\sqrt{2}$ secret. Consequently, in many of the stories in this book the interactions and dialogues between many of the math characters are often contentious. There are doubts, harsh challenges and animated questioning. Just as it is human nature to often resist change and revolutionary ideas, so the math characters are reticent to accept mathematical newcomers. Yet, in the end, challenges and questions yield to logical thought and proofs similar to how ideas are actually scrutinized in the world of mathematics.

In NUMBERS & OTHER MATH IDEAS COME ALIVE numbers and math ideas tell us their stories in their own words. We learn about their foibles and strengths. We see why mathematicians had to develop new concepts when the old ones were not sufficient for the problems at hand, and how math is constantly growing and evolving. We learn how new numbers appeared, and old ones had to accept the newcomers, and how certain math ideas lay hidden for centuries, until, they were resurrected or discovered by a mathematician. Love, hate, anger, excitement—all emotions we associate with humans are taken on by these math characters as their stories unfold. Hopefully you will be intrigued as you look behind the scenes at what numbers, points, lines, and other notions are saying and thinking. In each story, math properties and characteristics are uncovered and explained through the dialogues and actions of the math characters. In fact, you may forget you are actually learning mathematics as you get caught up in the characters' private lives.

numbers
&
other math ideas
come alive

Numberville

... number is merely the product of the mind. **—Karl Gauss**

Numberville is a town full of solutions. The population is infinite and only composed of numbers. It is where mathematicians look for solutions to the host of problems they are continually creating.

The recent election results in Numberville came as a surprise to numbers. Most thought there was no way the so called Reform Party would get elected, but by some quirky circumstances, the finite elite subset of whole numbers {0,1,2,3,4,5,6,7,8,9} were the only ones that voted. And, yes, you guessed it, the Reform Party was elected. Most of us did not realize what this meant. We numbers had become accustomed to the status quo, and were so into our own properties and factors that we had become oblivious to certain situations developing in Numberville. Apparently the numbers in this subset of whole numbers were accustomed to being one unit apart and did not like being crowded by all the new numbers that were arriving. First there were the proper fractions, who liked to socialize, and soon mixed numbers and improper fractions were all over town. Then the irrational numbers started arriving on geometric objects such as the sides of right triangles or diagonals of squares. As new and complicated problems developed, they provided many more solutions involving new numbers. Consequently, there was not as much work for whole numbers as there had been in the past. But this did not bother them because they received hefty royalties from the other numbers.

Apparently what these elite whole numbers objected to was the way these new numbers were hanging around town in no apparent order. Before the newcomers had arrived, they had always kept things neat and themselves in ascending and

"Just think of having to drag around an infinite tail of nonrepeating digits all day long!... "

descending order. This was important to them so that they could easily find a number that might be needed for a particular solution. Other numbers thought this was not necessary, and besides they did not want to be told where and near whom they always had to be.   Only the fractions and the irrational numbers suspected the real reason. The whole numbers were not skilled at distinguishing the size of numbers other than themselves. So the first thing that the Reform Party did was pass the ORDINANCE OF ORDER—*All numbers must display their decimal equivalents when in public places. No number could remain out of sequence for more than 10 minutes.*

Needless to say when this ordinance was posted, protests could be heard all across town. First the fractions and other rational numbers gathered signatures and presented their petition which claimed that the ordinance was an invasion of privacy. But the Reform Party simply said the petition was preposterous and rejected it.

The irrational numbers became very agitated by this arbitrary decision and decided to hold a protest. They held their demonstration in front of city hall holding up their radical signs in protest of the ordinance.

Once again the Reform Party members ignored the protests, and warned them that they were in violation of the ORDINANCE of ORDER and they threatened them with fines unless they displayed their decimal equivalents at this public gathering and arrange themselves in order.

With this declaration, a hush suddenly came over the irrational numbers and then whispering began to replace the silence.

"These whole numbers are so stupid," $\sqrt{2}$ whispered to π.

"They don't understand what they are asking for," π added very quietly.

"Don't they know what a burden this places on us. It's not so bad for fractions, even 1/7 has it easy with its decimal eqivalent $0.142857\overline{142857}...$* by using a bar over its repeating part," $\sqrt{2}$ said.

ANNOTATION_____

Every real number can be expressed as a unique decimal.

On the other hand, a rational number can be expressed in various number forms. For example, 7 can be written in infinitely different ways as a fraction: 7 = 7/1 or 14/2 or 21/3, etc., but 7 has only one unique decimal representation, 7.0. It is understood that infinitely many 0s are repeated after 7's decimal point without affecting its value. The decimal equivalent of _ is 0.75 but .75 is used for convenience, and again infinitely many zeros are understood to repeat after .75 without changing its value. On the other hand, 5/12 as a decimal repeats a pattern of numbers other than zeros, namely .416666... and to shorten its notation it is written as   , the bar meaning the 6 is constantly repeated. In the case of 1/7 we have . All rational numbers have a repeating decimal pattern in their decimal expression. However, any real number that is not a rational number (i.e. an irrational number) -meaning numbers that cannot be expressed as fractions whose numerators and denominators are integers-have decimal equivalents which have infinitely many decimals which never repeat a pattern.

"Just think of having to drag around an infinite tail of non-repeating digits all day long! It is cruel and unusual punishment," π loudly declared, "Especially since mathematicians have recently carried my decimals out to millions of places."

" Let's give them exactly what they want," e shouted.

"What do you mean? " $\sqrt{3}$ asked.

"They think things are crowded now. Just wait and see how crowded they will get," e whispered.

"Yes, Yes," other radicals joined in, realizing e's plan. Suddenly all the irrational numbers

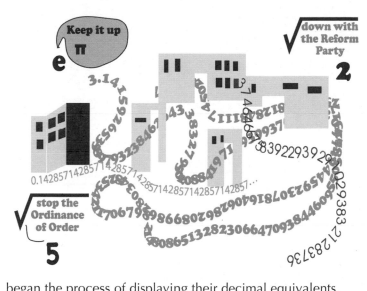

began the process of displaying their decimal equivalents.

The Reform Party whole numbers were aghast. The front of city hall was a mass of digits which had begun to fill every street, alley, stairway, room, and space. There was hardly any space left to move around. Numbers were pushing against one another. The whole numbers couldn't find one another, and were being scattered all over Numberville.

"Stop! stop!" the first ten whole numbers shouted. " We didn't know what havoc this would cause. We withdraw our ordinance. Just return things to the way they were." With that the irrational numbers promptly withdrew their decimals and donned their radicals and other symbols, and things returned to the way things were in Numberville.

The transcendentals

Pure mathematics is, in its way,
the poetry of logical ideas.
—Albert Einstein

"**Y**ou're just having an identity crisis," said e.

"How would you know? You're one of the most famous numbers, but me, nobody knows I exist. … I don't even have a name. …I'm just lumped in the general set of real numbers.
"All you other real numbers have names and symbols. Not me. People are not frightened by you. They come across you many times in their everyday work,… BUT how do they describe me?… as the solution to an non-algebraic equation.[1] What a mouthful! No wonder no one wants to come near me," the transcendental carried on, and couldn't stop complaining.

Finally, e had to shout to be heard, "Did you forget that I and π are also transcendental numbers?"

"What the heck does trans…transcen…transcenDENTAL mean? It sounds like something for teeth."

"No," e said laughing. "You, I and π and many many other numbers are transcendental. We are transcendental because were are non-algebraic."

"Now I'm being described by not being something. I don't even know what 'algebraic' means," transcendental lamented.

"It simply means there is no polynomial equation with integral coefficients that has me, you or π or any transcendental as a solution," e began to explain.

"Simply means? You've got to be kidding. Well, what about √2? Its decimal form has infinitely many non-repeating decimals, as I have, so what is its algebraic equation?" transcendental asked.

"That's easy to explain," π began. "It's polynomial equation is $x^2-2=0$. It's coefficients are the integers 1 and -2, and its solutions are √2 and -√2. Let's face it, neither you nor I can be expressed by such a polynomial equation."

"I know, but you and e have names, and most people associate you with the irrational numbers. Some people don't even know that you're transcendental," the transcendental continued on its tangent. "π and e are your actual names, and those names intrigue people."

"Stop complaining," π said in a quiet voice, trying to calm the hysterical transcendental, but it didn't help.

The transcendental wouldn't quit… "I'm going to end all this. Call it quits. I don't want to have anything to do with numbers any more. Why was Georg so stubborn? According to Georg, I am a member of the uncountable set."

"Georg? Georg who?" e asked.

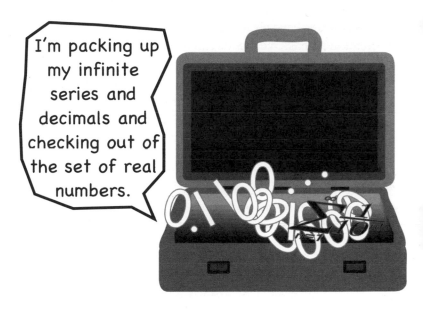

I'm packing up my infinite series and decimals and checking out of the set of real numbers.

"The mathematician Georg Cantor. He insisted on giving us, and that includes you, a public profile. Most of the conventional mathematicians didn't want to have anything to do with us. Since I am among the non-algebraic, that virtually makes me part of an unknown and unwanted set. Why did a mathematician even bother to discover me, let alone name us with such a scary sounding word, TRANSCENDENTAL?"

"It's not that bad," π said.

"Not bad? At least 99% of the people don't even know I exist. The whole numbers won't talk to me. They think I'm too weird looking, that I'm some sort of mutation. Why couldn't mathematicians have given me a short symbol name like you or even the √2 has. Neither of you have to carry your infinite tail of decimals all around.   I get no respect. I'm avoided like the plague. I'm packing up my infinite series and decimals and checking out of the set of real numbers." Transcendental was serious. "And you know what, I'm going to encourage all the other transcendentals to come along. You can join too π and e, if you want."

"Stop!" e shouted. "You can't do this."

"What do you mean?" transcendental asked.

"I mean you, I and all the other transcendentals are indispensable to the set of real numbers. If we leave, the real numbers will be decimated. We way outnumber all the integers and rational numbers. The reals depend on us — the irrationals and the transcendentals — for their substance. The real number line would have giant gaps in it," e explained.

"How do you know this?" transcendental asked.

"Didn't you know that your Georg Cantor proved this in 1874?" π answered.

"And in addition you were the very first transcendental that was specifically created to show that transcendentals were for real." e said.

"Really?" transcendental asked, somewhat astonished.

"Really!" e replied.

"I had no idea. That makes a big difference, and knowing we outnumber all the whole numbers, plus the integers, plus rationals...that's a big bonus. With this ammunition, I can handle the whole numbers." A big grin came across the transcendental's face as it began to unpack its series and decimals.

FOOTNOTES_____

[1] An algebraic equation is a polynomial equation whose coefficients are integers, e.g. $-3x4+7x=19-x^3$. The numbers $\sqrt{2}+1$ and $\sqrt{2}-1$ are solutions for the algebraic equation, $x^2-2x+3=0$.

A transcendental number is a solution of a non-algebraic equation such as $\pi x^2 +3x^4=5x+7$, $y=\log x$, $y=3^x$, or $y=\sin x$.

Some examples of transcendental numbers are the following where letter "a" represents any algebraic number except 0: $e, \pi, e^a, 2^{\sqrt{2}}$, sin a, cos a, tan a.

ANNOTATION_____

Leonhard Euler first believed that real numbers that were not algebraic existed, and named them transcendental because they transcended algebraic operations. Even though he believed in their existence, he had not come up with an example of one. It was not until 1874 that Joseph Liouville gave the first example of a transcendental number and proved it was a transcendental —

$$1/10^{1!} + 1/10^{2!} + 1/10^{3!} + \ldots = 0.1 + 0.01 + 0.000001 + \ldots =$$

$$0.110001000000000000000000100\ldots = \sum_{n=1}^{\infty} \frac{1}{10^{n!}}$$

The staunch group of traditional mathematicians of the 1800s felt these numbers were of little use and were few and far between. Imagine their surprise and concern when Cantor proved that the transcendentals are a larger infinite set than the algebraic numbers.

The two famous transcendental numbers π and e remained only irrational numbers until the late 1800s. In 1873, e was proven to be transcendental by the French mathematician Charles Hermite; and in 1882, π was finally shown to be a transcendental by the German mathematician Ferdinand Lindemann.

There is nothing negative about negative numbers

If people do not believe that mathematics is simple, it is only because they do not realize how complicated life is.
—John von Neumann

The integers were doing their daily calisthenics on the number line. zero was shouting out directions. "All right, you positive integers even up your spacing. That's better. –6 move a bit closer to –5, and 4 stay in line with the rest. That's it. Now, you look like the coordinates of a number line. Take a 5 minute break, and we'll resume our workout, and later we'll add the y-axis for a little variety."

The integers disassembled. Many gathered around the coffee bar getting an espresso or latte and relaxing.

"You know, I'm not sure I like being called a negative number," -6 said. "It sounds, well, so negative. I wish they had thought of something with a more…"

"Positive sound," +6 interjected.

"Exactly," -6 responded.

"Your negative doesn't mean anything negative," +6 continued, "it's just what those mathematicians came up with."

"Well, I wouldn't say it never had a negative connotation," said 1, the only number who knew everything from the past.

As the first number, it had been around and used by people ever since it was first scratched in the dirt.

"I remember times when negatives didn't even have a name, but just appeared in problems. They took mathematicians by surprise," 1 declared.

"We did?" -6 asked.

"Yes, you did!" 1 replied. "Not only did you surprise mathematicians, but, in fact, you startled many. It began when you appeared as solutions to problems. You were immediately tossed out and considered worthless answers.[1] Some felt that there was nothing beyond nothing, or should I say beyond zero."

"That's better," zero said in a hostile tone.

"It was just an unfortunate slip of the tongue. I apologize," 1 immediately added not wanting to start any discussions. "We all know your history zero, and how valuable you are." Over the years, 1 had learned how to flatter zero.

"Apology accepted," zero replied in a much more gracious tone.

"Now, don't fret about being called negative six," 1 said consoling -6. "Consider your origins and uses. Granted, the Greeks primarily used only positive quantities because their geometry did not lend itself to the concept of negative distance, area or volume. Poor Diophantus! When he came across negative roots in some of his equations, he would just choose to ignore negative answers since, to him, they were meaningless."

"But when did the negatives gain some respect?" -6 asked.

"Hindu mathematicians were the first to really work with negative numbers by doing arithmetic involving positive and negative numbers. This work enabled them to discover the

"Mathematicians still considered negative numbers untrustworthy, dubious, fictitious and ridiculous. "

two square roots of a positive number—one positive and one negative," 1 said, sounding very erudite.

"So the Hindus gave respect to the negatives?" -6 asked more confidently.

"Well, not exactly," 1 continued. "They were still wary of negative numbers, sometimes referring to negative roots as inadequate."

"Even when we appeared before the mathematicians' eyes, their minds couldn't accept us. They couldn't see our worth," -6 said sadly.

"Don't despair," 1 went on. "Your story doesn't end here. Negative numbers began appearing in everyday uses. The Chinese began using negatives in their record keeping. In fact, in the 12th century, the Chinese used red rods to designate positive values and black ones for negatives."

"Oh, I see. It was an early way to symbolize us," -5 noted.

"These everyday uses led to your present day symbol, '–.' Accountants and merchants started using the concept of negative and positive numbers in inventorying and taking measurements of containers.[2]

"Ah, respect at last," -6 signed.

"Not quite yet," 1 interjected once more. "Mathematicians still considered negative numbers untrustworthy, dubious, fictitious and ridiculous."

"What!" -6 shouted, very upset.

"Relax. Let me continue," 1 said trying to console and calm -6. "This didn't really change until the 16th century when Girolamo Cardano recognized the worth of negatives in solving quadratic, cubic and higher degree equations, and even square root problems. Although square roots of negatives surfaced in the work of ancient Greeks, their value was given no credence. It just took time, -6, for your sophisticated concept to sink into the heads of mathematicians. And if they hadn't accepted you, mathematics would have remained at an impasse. For example, look at the simple quadratic equation $x^2-36=0$. It would only have one solution, namely 6, if your idea were not accepted. In fact, the whole elegant theory about the number of roots of an equation being tied to the equation's degree would never have been discovered. Because you, -6, are the second solution for this equation, and you and 6 are its only solutions…exactly two solutions just as its degree indicates it must have."

"Stop, stop, stop!" shouted zero. "Why are we continuing to discuss the past life of negatives? Today the term negative in mathematics is on firm ground. Not only accepted and used by all mathematicians—but laypeople as well." zero's voice rose to a crescendo, concluding— "Every new number has gone through similar identity crises and criticisms. Today, there is nothing negative about negatives!"

With that comment, -6 straightened its shoulders and walked away proudly.

---

[1] Although in 1484 Nicolas Chuquet's *Triparty en la sciences des nombres* was the first algebra book to use negative numbers, he could not accept them as solutions to equations. In the 16th century mathematician Michael Stifel author of *Arithmetica integra*, which summarized the known arithmetic and algebra, referred to negative numbers as absurd.

[2] Containers' measurements were gauged above and below a fixed amount or weight. It was here that the symbol, '–', appeared.

When the new domains came into the picture

Mathematics is the gate and key to all sciences. —**Roger Bacon**

"**Y**ou know," 1 said, "our domain isn't what it used to be. Numbers have changed." 1, the oldest number, was chatting with its closest countable friends 2 and 3.

"I know exactly how you feel 1," 2 began. "In the very old days our domains were less complicated, and the ranges for solutions to problems were always found in the whole number."

"Right!" 3 chimed in. "People didn't have to contend with fractions. Things were a lot simpler. Solutions were neater. What's our world of numbers come to?" 3 asked.

"A population explosion and complicated functions—that's what! There are more numbers than ever, and I'm not just talking about an infinite number of numbers. I'm talking about the many types of numbers that have been invented or should I say popped up as solutions to the ever crazier problems and functions that mathematicians try to solve." 1 was now ranting. "Remember when we were the only show in town? We supplied the counting numbers, and they were all that people needed to use in those early days."

"Yeah! Yeah!" 2 shouted. "I even remember when people just used the ten fingers of their hands to figure out problems. And later on, people began doing their figuring by pencil or in their heads. Then problems and their answers got harder and more complicated."

"Really?" asked 3.

"Yeah, people had to begin working with the decimals and fractions in problems. And a few mathematicians even began estimating the value of π out to hundreds of places and making complicated tables for radical numbers, trig functions and logs," 1 explained.

"What's happening?" $\sqrt{7}$ *walked over and nudged itself between 2 and 3.*

"I recall when that began to happen," 2 said. "I remember far enough back when the only numbers were discrete like us."

"Shhh…here comes one of non-discrete ones," 1 warned.

"Hi !" $\sqrt{7}$ said. "What's happening?" $\sqrt{7}$ walked over and nudged itself between 2 and 3.

"What's the big idea of butting in?" 2 demanded angrily.

"Why are you upset, 2? This is where I belong in the order of things." $\sqrt{7}$ replied.

"We were just chewing the fat," 1 said in an attempt to calm down 2.

"Speaking of fat," √7 injected, "you all look like you're putting on weight. Not much work these days, heh?" √7 teased. "Haven't seen you called in on many problems lately. In fact, it seems you've all been either on a leave of absence or temporarily laid off."

"What do you mean fat? laid off? We are as important and useful as ever," 2 said, its voice showing how annoyed it was. "And, in case you haven't noticed, I've been called in on a lot of work with odd and even expressions, and prime number stuff. Moreover, we were just saying at least we're among the discrete numbers, unlike you and your friends."

"Really?" √7 smiled. "So we're not discrete?"

"That's right," 1 answered. "There's no other counting number between any two consecutive counting numbers, and the same can be said for all the integers. But, when it comes to radicals, fractions, decimals, and transcendentals, there's always another one to be found between any two of you.. That makes you dense in more ways than one," 1 said chuckling.

√7 was mad and shouted, "Now just one minute, we may be dense and not discrete, but we are not elitists like all of you. In fact, we're very inclusive. I'm not just standing between you, 2 and 3, by coincidence. It's where I belong, and you're certainly not making me welcome."

π, overhearing this heated discussion, walked over to the group. "Now, now no reason to raise your voices," π said addressing 1, 2, 3 and $\sqrt{7}$. "What seems to be the problem?" It asked.

"You're one of the transcendentals," 1 declared. "There's no way you can help."

"I overheard you 1, 2, and 3 referring to yourselves as discrete numbers, but actually we are all part of what makes the real numbers a set that is not discrete," π began to explain.

"What are you saying?" 3 demanded.

"Sure, I agree that the set of counting numbers are discrete unto themselves, but still you are part of the much greater set, the set of real numbers. Correct?" π asked.

""That is true," 1 replied.

π continued, "Discrete doesn't mean you're set apart from the reals. You are as much real as I am and the $\sqrt{7}$. And as real numbers we are all part of the same family, or should I say same set, the set of real numbers."

"We never thought of it like that," 1, 2, 3, and $\sqrt{7}$ replied in unison.

"Furthermore, without one of us, the real number line, which happens to be expressed by the function $f(x)=0$, would have a gap, a missing coordinate, for one of its points. Agreed?" π asked.

"True," the other numbers answered.

"And although as real numbers we are no longer considered a

discrete set of numbers, we ARE more importantly part of the continuous function f(x)=0. One for all and all for one real number line! Each of us is essential for supplying each and every point of the real number line a numerical name. So stop your teasing and arguing and get to your particular point and name your coordinate."

1, 2, and 3 no longer felt so crowded or imposed upon by all the other types of numbers between them on the real number line because they realized without each of the counting numbers and their respective points, the real number line would look like a dashed line.

Not all symbols are created equal

So if a man's wit be wandering, let him study the mathematics; for in demonstrations, if his wit be called away never so little, he must begin again. —Francis Bacon

"**F**rom their inception, numbers knew they were not all equal to one another. In the beginning numbers were only used for counting things. But slowly the operators began to creep into their everyday lives.

Addition and subtraction were the first operators to begin working with numbers. They were naturals since people were giving and taking things back and forth all the time. And, as larger and larger quantities began to be used, it just wasn't physically feasible for humans to carry and exchange large quantities back and forth to solve problems. At some point someone figured out a few short cuts and devised the operators multiplying and dividing. More and more numbers of all sizes were now being involved in everyday problems by using +, -, x, ÷. Jobs for numbers of all sizes were in demand. The equal sign seemed happy being busy as it had never been before. It felt alive, important. Using the operators, it could now settle disputes, put things right, keep harmony. But it didn't take long before the numbers came to the realization that equal was not enough. There had to be more to life than always making sure things were perfectly balanced.

Equal tried its best to keep the numbers occupied with all sorts of problems and games. Both humans and numbers realized equal was overwhelmed trying to persuade people to operate with numbers so that things would always come out even. But humans kept coming up with new problems. Problems that were just not fair. Humans and numbers were no longer content with equality. They continually wanted more, more and more. But, some would get too greedy, and end up with less, less and less. Humans were bored with always having to use operators to end up with things being equal. Equal became a very tedious symbol to both numbers and people.

"Humans thought they had problems solving equations, just wait until they begin using these symbols..."

Equal realized the inevitable. There was no way to set 5=3 without adding 2 to 3 or subtracting 2 from 5. They just weren't on equal footing. Equal struggled to convince, cajole and con the numbers to use operators. But 3 and 5 would not budge. Equal was beside itself. 5 shouted, "We are tired of using operators and you to relate us."

"There has got to be another way to connect us," 3 insisted.

"Do something else, and break this monotony!" 3 and 5 demanded.

So equal's two equidistant bars began to squirm, twist, pull, bend, and finally = became >. Equal shouted "five is greater than 3 shall be written 5>3."

"If it is so" yelled 3 and 5, "then we realize we can actually be related without resorting to adding or subtracting or other operations."

But the next day 3 urged and dragged his big cousin 5 to go and see equal. "We like the new way we're related or should I say connected, but why do I always have to be behind 5?"

"Besides," said 3, "5 pushes a lot."

Equal was stymied. "What do I do now?" it thought. "These numbers and their egos were sometimes too much to deal with."

"Change positions," equal finally said. So 3 stood to the left of 5, and equal pulled on itself no longer leaving the constant distance between its bars... and created 3<5. "This will now mean 3 less than 5," equal replied, exhausted from the strain.

But equal did not stop here. "I don't need to be the only symbol relating numbers. I need help with all these numbers and problems. Go to the other numbers, and let the word be known that there are many other ways they can be related."

"We already told them about ">", and now we will also explain "<" to them ." 3 replied.

" Don't rush off too quickly," equal shouted.

"What do you want?" 3 asked.

"Your new connectors don't stop with > and <. I have also worked on others?" equal said.

"Others?" 5 asked.

" I have made $\geq, \leq, \neq, \geq, \leq, \approx$, and there may be even more." equal replied.

 "Wait! Stop!" interrupted 3 & 5. "What are all these?"

"You'll find out soon enough," equal said with a funny grin on its face.   "Humans thought they had problems solving equations, just wait until they begin using these symbols...for not all symbols were created equal." Equal nodded slowly walking away smiling, leaving 3 & 5 connected by < and thinking and laughing, "Free at last. Now I can relax, and let the humans deal with putting together the numbers the way they want."

The day an isosceles triangle made the golden rectangle eat crow

Mathematics is an escape from reality.
—Stanislaw Ulam

Everywhere you looked there were triangles. Large, small, fat, thin...dashing around bumping into each other's vertices. The plane looked as if it were a sea of triangles. That is until a rectangle seemed to appear out of the blue, not any rectangle, but the golden rectangle. And, as if by command, all the triangles suddenly did an about face in the direction of the rectangle.

"So how's it going? " golden rectangle shouted with a superior tone to its voice.

"How's what going?" a triangle asked. "My three sides are doing well, my three angles still total 180, and I'm still an isosceles triangle."

" So nothing to complain about," golden rectangle said. "Same old, same old, eh? You look as boring as usual," golden rectangle added with a chuckle. " I'm sure glad I have 4 sides, 4 right angles, 4 vertices, and two congruent diagonals that bi-

sect each other. Yes, I am even
more exceptional, since I'm a
very special rectangle. I can
generate an equilateral spiral just
by drawing squares inside myself
like this."

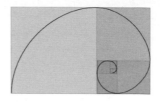

"I have heard this from you so many times before. The same
old gloating comments. The first time it was interesting. In
fact, I looked up your golden properties, and yes they make
you stand out from the other rectangles. But triangles are even
more distinctive than rectangles" the triangle replied boldly.
"We have so many other ways of distinguishing our various
types of triangles. Rectangles other than the golden rectangle
are just all called rectangles," the isosceles triangle explained.

"Now just one minute," golden rectangle interrupted.

"Now just one minute yourself," the isosceles triangle retorted.
"You never bother to listen and look at us triangles. You think
you're so great."

"But you all look the same to me," golden rectangle
countered.

"Have you had your eyes checked lately?" isosceles triangle
teased.   "Look!" isosceles triangle pointed, as it continued to
comment. "There are scalene triangles, obtuse triangles, acute
triangles, right triangles, equilateral and equiangular triangles,
right isosceles triangles, acute and obtuse isosceles triangles
all around us."

Rectangle was taken aback by isosceles triangle's bold statements.

"Come off it. Face it! You are just getting touchy and old. In fact you triangles are the oldest of convex geometric figures. So get over it. You may have lots of named types of triangles, but I remain the golden of my species and all geometric objects," golden rectangle boasted.

"I have nothing to get over. I have self-awareness. I know what I am, and how useful I am and have been to mathematicians. All the triangles you see around you have subsets of characteristics," isosceles triangle said loudly.

Golden rectangle looked around, and realized he was surrounded by thousands of triangles, all pointed at it like arrows.

 Golden rectangle decided to listen a bit to what this outspoken isosceles triangle had to say, especially since it felt vulnerable with all these arrows pointing at it.

"You're saying your vertices are always on the same plane?" golden rectangle asked somewhat timidly.

"That's true. Our three vertices determine one plane unlike your vertices," isosceles triangle explained.

"Now just a minute," golden rectangle interjected. "I am a planar object, so I am on one plane."

"True," isosceles triangle said, "but your are on one plane because that's how you are described, or should I say defined. My vertices *determine* a single plane. In other words we define the plane we're on, not the other way around."

"Wellllllll" golden rectangle uttered. "That may be so, but remember none of you can be golden with its own equiangular spiral defined within it," golden rectangle said smugly. Feeling very satisfied with its reply, it began to walk away.

"Hold on there. I guess you don't know about how special an isosceles triangle I am because I don't go around all day bragging about it."

"Ha, ha, ha…what a good joke," golden rectangle snickered. "What's so special about your two congruent angles and two congruent sides?"

"Do you see my base angles are each 72° and my vertex is 36°?" isosceles triangle asked.

"Yeah," golden rectangle replied.

"What happens when one of my 72° base angles is bisected?" isosceles triangle asked the golden rectangle.

"Well, it's cut in half and becomes 36°. So?" golden rectangle questioned.

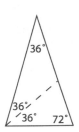

"Now look closely at the bottom small triangle with the 36° and 72° angle. Do you notice

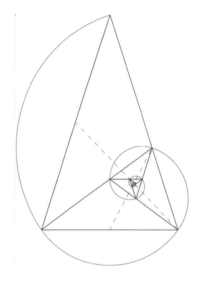

something?" isosceles triangle asked.

The golden rectangle looked and looked and thought and thought. All of a sudden it said "Oh my gosh! That's another smaller 72°-72°-36° isosceles triangle."

"Right! and when you repeat the process—like the process with inner squares within you—my inner isosceles 72°-72°-36° triangles' vertices also outline an equiangular spiral, just like yours. I am called a GOLDEN TRIANGLE!" isosceles triangle said.

Golden rectangle blushed. It felt so foolish it dashed away.

Isosceles triangle turned to its triangle friends with a huge grin on its face. The scalene triangle came forward and patted the isosceles triangle on its vertex angle saying, "I guess you finally put an end to the golden rectangle's continual boasting. I guess the golden rectangle was not as bright as it thought it was." With that you could hear a roar of laughter erupt from the sea of triangles.

Little wavelet shows off its stuff

Where the telescope ends, the microscope begins. Who is to say of the two, which has the grander view? —**Victor Hugo**

"**M**ove it, shrimp", shouted sinus wave, appearing huge next to the minute wavelet. "How can a puny curve like you make waves?" sinus curve asked, taunting the little wavelet. "You're not even big enough to make a ripple," sinus wave continued with a snicker.

Whenever poor little wavelet* came near any of the normal sized sinus curves it knew it was in store for some heavy duty teasing and today was no exception. No matter how big or small the other sinus curves' periods and amplitudes were, wavelet could not hold its own next to them, let alone hardly be visible. The large sinus wave continued to belittle wavelet saying, "How could a mathematician use such a tiny curve as you for anything?"

"I know I'm used in solving problems and even describing sound," wavelet replied trying to sound confident and unintimidated by their comments.

"Just think about it, what sound would be so short to be described by you, scrawny?" a really big sinus wave countered in a deep husky voice.

Wavelet was visibly upset as it skittered around. Angered by these words it finally replied, " You might be able to describe the sound in a room with your big wavy shape, but you devour all the individual noises. Sounds can lose their identities in the shape of a wave, and your function's equation doesn't help distinguish or reveal the individual noises that make up the sound in this area."

"What do you mean?" shrimp, big sinus asked, confused by wavelet's statement.

"You and the other sinus curves are like cake batter!" wavelet said.

"Cake batter? You're not only puny in size, but also in mind," big sinus said as all the rest laughed.

Wavelet felt really bad, but nevertheless it began to explain its comment. "Yes, you're like cake batter! Take sugar, eggs, baking powder, flour, milk, seasonings, and butte. Mix them together and what do you have? ...batter. All the individual ingredients become disguised in the batter. Take the sound your sinus curve produces. It's just like batter. It is impossible to distinguish all the individual components that compose it. BUT, if you take a conglomeration of wavelets, one can zero in on the component sounds. We, wavelets, let mathematicians pick out sounds. We are invaluable!"

"Have you gone bonkers? We sinus waves have never heard of such things," big sinus replied.

"I can't help it if you all are ignorant." Little wavelet was giving it back.

"Ignorant? Better watch what you say wavelet, or we may have to smash into you," big sinus curve said as it pushed wavelet.

Wavelet pushed back saying, "Look, let me educate you all. Each wavelet represents a tiny interval of time, called a

"You and the other sinus curves are like cake batter!"

window*. In other words, during this window of time you hear the sound of that minute interval, which a wavelet describes with its shape and equation."

"Just a moment," big sinus interrupted. "You're no different than cake batter either. Just because you cover a tiny increment of time doesn't mean you haven't swallowed up a sound shorter than you."

Sinus wave and wavelet were at an impasse. Both were right.

Wavelet began to feel ill, secretly admitting to itself that it was a blend of minute sounds.

"Wait," a very small voice yelled, "I am the answer"

"Who said that?" all the sounding waves asked as they scanned the area around themselves.

"I did. I am a distinct tiny wavelet identifying an INDIVIDUAL sound because my shape and equation are not determined by a fixed interval, no matter how small or long. Instead I am determined by the PITCH of sound. The sound determines me, not some time interval."

"Who are you? Where do you come from?" All waves and wavelets asked simultaneously.

"I am a Daubechies wavelet*. I came from the same place from which you originated....the imagination of a mathematician. You can think of me as an EVOLVED sinus wavelet. Don't feel bad, you all will always be useful and important concepts, it's just that I'm more distinct because each sound retains its individuality when I'm used." Wanting to appease the waves Daubechies wavelet added, "You guys know sinus waves are invaluable. We sinus curves—and I do mean WE—are used to describe lots of things other than music and sounds."

"Like what?" sinus wave questioned.

"Like all sorts of data, " the Daubechies wavelet answered.

"What exactly do you mean?" wavelet asked.

"We are used to compress data," Daubechies wavelet replied.

"Compress data...what does that mean?" sinus wave asked.

"Look," the petite voice of the Daubechies wavelet began to explain. "Data is information which comes in all forms such as sounds, pictures, photos, shapes. This data is rewritten using formulas composed of such equations as our functions."

"But why does data need to be compacted?" sinus curve asked.

"So we don't have to use a lot of space to store the data, and more importantly, to quickly transmit it electronically. These compact equations, often referred to as functions, are written along with a set of values which are used to convert a formula back to its original form, be it music, a photo, a fingerprint, or what ever it was. It's almost like writing a secret message and then using the code to translate it back," Daubechies wavelet elaborated. "The important thing to remember is we are ALL important in the process. We all make an impact. We all make waves."

"We ALL make waves!" all the waves shouted their mantra together.

Even though wave theory dates back to the early 1800s with the work of French mathematician Jean Baptiste Fourier, a wavelet is a relatively new mathematical concept.   Wavelets can be used to describe anything that appears as a picture, sound, vibration or other form of data.   A **wavelet** is a certain type of mathematical function which divides a given function or continuous-time signal into different frequency components so that each component can be studied with a resolution that matches its scale. The time component of a wavelet is called a **window**. Its size can vary according to the scale or resolution being used.   When a combination of wavelets is used to describe a function, it is known as a **wavelet transform.** The wavelet transform converts a wave or signal into a series of wavelets in which each wavelet preserve its characteristics, while describing the wave or signal being transformed into its small wavelet components. Wavelet transforms have many applications and are used in various fields. Among these are: physics, astrophysics, seismic geophysics, optics, turbulence, quantum mechanics,   blood-pressure, heart-rate, ECG analyses, DNA analysis, protein analysis, climatology, fingerprinting, general signal processing, speech recognition, computer graphics, multifractal analysis, and image processing.

A major breakthrough in wave theory came in 1987 with the wavelets devised by Ingrid Daubechies. Daubechies introduced time in a new way into the graph of the waves. She connected the pitch of the sound to time. The higher the pitch the shorter the wavelet. Thus, each wavelet describing each sound is a single curve unto itself, yet a string of wavelets can be used to describe a conglomeration of sounds or objects. These wavelets are like fractals of sound—the further away you stand from the wave the general shape of the sound of the room appears, but as you zero in on the wave the individual wavelets come into view. The Fourier mini waves are periodic, they repeat their shapes. Yet, Daubechies' wavelets are not repetitious.

Math tailors the numbers

A mathematician is a blind man in a dark room looking for a black cat which isn't there.

— Charles Darwin

If you think people are vain, you haven't seen anything. Some numbers are very picky about their wardrobes. In fact there was a time when integers had to wrestle with the square root of 2 and its decimal expression. The square root of 2 dates back millennia to the Babylonians*. Later its decimal approximation began appearing, and these decimals were a struggle for centuries until…

* * *

"Stop fussing with your decimals and adding more digits to your train," 3 shouted to the square root of 2. "Your train is long enough."

"If you've got an infinite number of never ending non-repeating decimals, why not flaunt them." Square root of 2 replied while arranging a 3 after the its 171.

3 wasn't the only numeral tired of square root of 2's antics. In fact all the integers were fed up not just with the square root of 2 but with the 100 zeros being swished around by a googol, and never ending non-repeating digits being produced by e's limit formula,

$$\lim_{n \to \infty} \left(1 + \tfrac{1}{n}\right)^n \approx 2.7182818\ldots$$

"We have got to do something" 3 said to 2. "Can't you speak to the square root of 2? After all you are somewhat related."

"I just don't feel close to it or have anything in common except the numeral two." 2 replied.

"Look, I don't go around bumping into every numeral with my tail of infinite zeros. You know I could be written as 3.00000… in decimal form. But what's the point? And why crowd things? Why can't the square root of 2 just use a short approximation?" 3 countered. "Even repeating decimals have given up the trains of digits. Remember when 1/7 was going around with its repeating decimal 0.142857142875142875… until it adopted the bar over the repeating part 0.1̄42875̄ just as 1/3 changed from 0.3333… to 0.3̄ and 5/6 went from 0.83333… to 0.83̄ . That bar the mathematicians came up with was a life saver."

"I still can't understand the square root of 2's behavior. Furthermore, the square root of 2 isn't the only irrational number traipsing around with its infinite non-repeating decimals. What about the square root of 3, 5, 6, 7, 8, 10, 11,…? They are all into this new dress fad," 2 added.

"Tell me about it. Have you seen what the golden ratio is doing? Even e and π sometimes get into the act. And what about the googolplex with its googol of zeros following its 1. It's out of the question!" 3 said and added "It's time to take some action!"

"But what shall we do?" asked 4 coming forward from the set of integers.

"Look, we all have nice neat simple numeral names that we wear. Right?" 3 said.

"Right!" all the integers replied.

"Let's start a fad that does away with low slung trains of numerals, stressing how difficult it is to walk with those outfits. And besides it's downright impossible for us to move around one of their never ending decimal expressions or the thousands and thousands of zeros being dragged around by googolplex. Any ideas?" 3 asked.

1 stepped forward and said, "Look at me. I have the simplest looking numeral. Just a straight segment. I think we should look to simplicity and the tailored look for a new chic style."

"Great idea, but how do we make this fad catch on?" 2 asked.

"It's all in the packaging," 5 spoke up. "Why not let all the irrational square roots be given a numeral name with a fashionable hat which will have tucked in it their never-ending non-repeating decimals."

"Hat? Do you really think we can pull this off?" 3 asked.

"No problem. For example the square root of 2 will be now written as $\sqrt{2}$. Isn't that a charming addition to its wardrobe?" 5 asked.

"It is cute," 4 said. "We may just be able to pull this off. But what about the cube root of 2? What hat could we devise."

"Why not something logical, like $\sqrt[3]{2}$ where the little 3 shows you are taking the cube root." 5 said.

"Very clever," 2 replied.

"Actually I can't take credit for it. It's the mathematicians and mathematics that devised this ingenious method," 5 explained. "In fact, to be consistent the $\sqrt{2}$ can also be written $\sqrt[2]{2}$."

"That makes all the sense in the world, and these hats are quite attractive," 3 added.

"Well, while we're at it I might as well show you another new outfit for the square root of 2. It's always good to give a choice, in case it doesn't go for the hat. So here's a complete make over $2^{\frac{1}{2}}$. Now isn't that fetching?" 5 asked.

"It's different," the integers replied.

"And for $\sqrt[3]{2}$ $\sqrt[4]{2}$ $\sqrt[5]{2}$ .... and on and on we have

$2^{\frac{1}{3}}$ $2^{\frac{1}{4}}$ $2^{\frac{1}{5}}$ ... " 5 demonstrated.

***

5 went off to meet the square root of 2 with two beautifully wrapped boxes.

"What's in those boxes? Candies?" square root of 2 asked.

"Oh you mean these. They're the latest numeral styles in hats and clothes for square roots and all other roots," 5 replied.

"Let's have a look," square root of 2 asked.

"Oh, I don't think you'd like to see these. And besides, I know you would not want to replace your long, heavy, awkward train of decimals with these fetching new numeral garments," 5 replied coyly.

"Have you seen what the golden ratio is doing?"

"No, I mean… yes I would like to see them. Who knows?   I may even like them." Square root of 2 said, its curiosity piqued.

"Well… they were not designed for just any numbers, but with square roots and other roots in mind. They have a tailored chic style to them." 5 laid it on thick, tantalizing the square root of 2's ego.

"Please stop and let me have a look." Square root of 2 was now pleading to see the garments.

"Since you insist, and besides I don't see any harm in letting you have a peek," 5 said as it began to unwrap the boxes.

"Oh my! They are gorgeous, but I especially love the hat. May I try it on," square root of 2 asked.

"Of course, but you will have to tuck your decimal train into it," 5 explained.

"No problem," square root of 2 replied as its decimal digits disappeared under the hat. "It's perfect. What do you think?"

"I think it's you, especially since it was designed with you in mind," 5 said with a satisfied smile on its face. "But you can only have it under the condition that you will share it with all the square roots and the other roots. You will make a fashion statement only if you all agree to wear them together."

"Of course we will. We'll get rid of those dreadful trains of digits. Googol and googolplex will be so jealous." $\sqrt{2}$ said, and ran off to share with the other roots.

"What about googol and googolpex?" 4 asked 5.

"Mathematics has also taken care of them. Just have a look at their slick new numeral outfits." 5 said holding up 10^{100} for googol and $10^{10^{100}}$ for googolplex.

"Clever use of exponents to cut them down to size. They'll love them," all the integers synonymously said laughing as the trains of digits emanating from the various roots of numbers began to recede into the new "hats" the irrational roots now wore.

*ANNOTATION_____

The concept of a square root first made its appearance on an ancient Babylonia cuneiform tablet—YBC7289 from the Yale collection. This Babylonian tablet was made sometime between the 19th and 17th centuries BCE. The tablet shows the √2 written in sexagesimal number form—

These Babylonian numerals represent 1, 24, 51, 10. The ⅄=1 and the symbol ◄ =10. By assuming the sexagesimal point is between 1 and 24, the number converts to the following:

$1 + (24/60) + (51/60^2) + (10/60^3)$

$$= 1+(2/5)+(51/3600)+(1/216000) \approx 1.41421\overline{296}$$

which compares to the √2=1.414213562... To arrive at their estimate the Babylonians probably used a repetitive approximation method which was often used by the ancient Greeks.

The actual square root symbol, √ , first appeared in print in 1525 in the algebra/arithmetic book, *Die Coss*, by German mathematician Christoff Rudolff. Some feel Rudolff's symbol originated from the letter, "r" in the word *radix*, the Latin word meaning root. In addition, Rudolff used c√ to denote cube root, and √√ for fourth root. Some contend the square root symbol originated with Arabic mathematicians and comes from the Arabic word, *jadhir, also* meaning "root".

Numbers & numerals try to set the record straight

We could use up two Eternities in learning all that is to be learned about our own world and the thousands of nations that have arisen and flourished and vanished from it. Mathematics alone would occupy me eight million years.

— Mark Twain

"**W**hy can't they ever get us straight," numeral wondered.

" People don't seem to be able to tell us apart. They don't know how to make a distinction between us," number replied. " At times, I feel that you and I are like Siamese twins, somehow joined at the number."

"I don't like it! We are different." numeral said, sounding very irritated. "It seems people have both a hard time telling us

"What do you mean, many symbols for the same number? I thought there was just one of me for one of you"

apart and a hard time talking about one of us without mixing us up with the other."

"What's so hard about getting us straight?" number questioned. "We have totally different personalities and characteristics. And remember, I came first," number said.

This comment irked numeral. " I know, I know," numeral lamented. "But, when people wanted to express a number, the only way they could communicate was to use a repetitive mark to show that particular number of things, and that's where I came into the picture. You were the idea or concept, and I provided the various symbols to name the different numbers," numeral expounded.

"Right!" number exclaimed. "As people began using various numbers in their everyday activities, they invented different symbols for the different numbers they used."

"Maybe they confused us because there are infinitely many numbers, each requiring a symbol and sometimes different symbols are made up for the same number. The symbol for the number 'five' is not unique, while its number quantity is," number said trying to bug numeral once more.

" What do you mean, many symbols for the same number? I thought there was just one of me for one of you," numeral said indignantly.

" Sorry to break this news to you, but no. There's just one of me and lots of you for each different number. See, a number is a specific amount while a numeral is a symbol for that amount," number explained.

"But the numeral for five is 5. Right?" numeral asked.

 "Well, 5 is one possible symbol for five. In China its symbol can be ||||| or 五 . In Roman numerals it's V, in ancient Egyptian it was ‖‖ and the Babylonians used ⟨⟨⟨ . See what I mean?" asked number.

" Now I can see why people get us confused," numeral replied.

" Come on, it's not that complicated," number insisted. "For example, if someone says pick a 'number' between one and ten, and seven is picked, they mean the amount seven not a numeral for seven as 7 or ⊬⊬ // or VII to name a few. But, on the other hand, if someone asks specifically to pick a number from 2, 4, 5, 7, 62, 123, they probably mean 'your favorite' from this group, since there is no quantity in mind."

"Huh?" numeral said still feeling very confused. "That does not make things clearer for me. I think you should have asked 'what shaped symbol from this group do you prefer?', because when you said number you might have meant its quantity rather than its symbol. See even you get things mixed up, number." numeral chuckled.

"I guess I can't blame people for getting us mixed up, when even we get confused sometimes. The thing to remember is that we are DIFFERENT," number emphasized as it bid numeral goodby.

What do you mean undefined?

The tantalizing and compelling pursuit of mathematical problems offers mental absorption, peace of mind amid endless challenges ... and the sort of beauty changeless mountains present...

—Morris Kline

In the land of space, stories were circulating about things that were not really defined. One day space, which is defined as the set of all points, called together all its points for a symposium. But, when all points get together you also have lines, rays, angles, planes, half-planes, squares, triangles, curves, dodecagons, hexagons, trapezoids, parallelepipeds, hexahedra, and on and on. In fact, new shapes appearing from points seem to never cease, and it seems mathematicians are always coming up with new objects.

There was quite a ruckus as they gathered in the vastness of space. Finally, when space had their attention, it said in a very deep voice, "Who's been spreading rumors?"

"Rumors?" asked triangle.

"Yes! rumors. There are stories going around about the undefined ones," space replied.

"Undefined ones?" questioned triangle.

"Yes! And why do you keep repeating what I say?" demanded space.

"Because…because I want to be sure I heard you correctly," said triangle.

"Well, you did," space replied.

"What do these rumors talk about?" asked the ordered triplet (2,-3,7), which was sitting on its point near space's x,y,z axes.

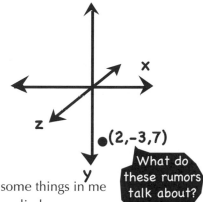

What do these rumors talk about?

"They imply that there are some things in me that are undefined," space replied.

"Undefined?" the x, y, and z axes asked in unison.

"You see," space said, "if there is something undefined within me I figure that would mean I'm undefined. Right?" it asked.

"Right," said the tetrahedron.

"Right, Right, Right,..." reverberated throughout space.

Suddenly among all the 'rights' a 'wrong' was heard.

"Who said that?" demanded space. "Who's questioning my logic?"

From a remote part of space a voice belonging to a point said, "I am."

"You are just a point. What do you know?" a trapezoid

snapped. "You're without dimension. In fact, you're basically invisible. You just show a location. How can you say space's logic is wrong?"

"Because I and all points are undefined," the little voice replied confidently.

"Look point, didn't I just describe you by saying you have no dimension? You just show location."

"Right you just *described* me, but you did *not* define me. There is a big difference," the point explained. "If I say you're a 2-dimensional object with four sides—that's part of your description not your definition. For example, a triangle is *defined* as the union of three non-collinear points and the points between them," point continued. "That's the mathematical definition of a triangle, but a triangle can also be *described* as a 3-sided closed convex figure."

"Yes, that's true," triangle replied.

"I am sure you will also agree that each definition consists of words. Right?" point asked.

"Right," replied triangle.

"And must each of those words used in the definition be a defined word?" asked point.

"Yes," replied space. "That is what I am getting at. I am undefined if any part of me is. Right?"

"WRONG!" point boldly replied. "It is *impossible* to define every word or object by using words and objects which are themselves *all* defined."

"What?" all the other objects in space shouted. "What do you mean?"

"That's right. They start with a few undefined words, objects and axioms and create an amazing imaginary world ..."

"I'll try to explain. When creating a mathematical system, mathematicians have to start with some objects or words which you just accept. These become building blocks for the mathematical system. New objects or definitions are created and defined using the building blocks because you just cannot make up things from nothing. In other words, to define something we use words, and some of these words will probably be undefined because you need to start with something to get something, otherwise you'll have nothing."

"Well...I guess so," triangle replied with a baffled look on its face.

"Look, for a triangle the undefined terms used are *point* and *line* along with the defined terms *segment, union* and *betweeness*. You've got to have a gut feeling what *point* and *line* mean. Then using the concept of a *point* and *line* we can define the concepts of *betweeness, union,* and *segment*. Once we have done that we can finally define you, triangle*." Point continued trying to clarify its thoughts. "You have to accept that some things are undefined and build with them," point smiled as it added this comment.

"So I suppose now you're saying *you* are a building block of geometry, " space grunted.

"Not just me but infinitely many things which occupy space," point said.

"What and where?" shouted space.

"Here, there, everywhere," point pointed out. "All the points, all the lines, all the planes. They are all also undefined terms."

Shocked looks crossed the faces of every point, line and plane of space. There was total silence.

"Where did you get this information?" space asked, snickering.

"From the imagination of mathematicians. Look at Euclidean geometry. They created this geometry," point explained.

"You mean I and all of you are figments of their imaginations?" space asked.

"That's right. They start with a few undefined words, objects and axioms and created an amazing imaginary world ...a mathematical system that makes sense, and somehow works in the real world. Some of us are chosen to be the undefined ones. We are all essential parts of this system. You, space, are defined. So are square, rectangle, ordered pair, circle, quadrilateral...The most important thing is that we are logical, we make sense, we..."

Space interrupted point exclaiming, "I GET THE POINT."

"Yes," said square. "yes, yes, yes..." spread throughout space as all its objects chimed in.

---

*ANNOTATION_____

The evolution of the definition of a triangle

The concept of *betweeness* of points relies on the undefined terms of *point* and *line*. Point B is *between* points A and C if the three points lie on the same line and |AB|+|BC|=|AC|.

The *union* of sets of objects is defined as the joining together of sets into a common set. Note, duplicate objects are not repeated in the union, e.g. {1,2,3} ∪ {2,4,6}={1,2,3,4,6}

A *segment* is defined as the union of two points and all points *between* those two points.

Now, using the undefined terms *point* and *line* and the defined term *segment*, the term *triangle* can be defined as follows: If points A, B and C do not lie on the same *line*, then *union* of the segments AB, BC, and AC form *triangle ABC*.

A square is not so square

Why sometimes I've believed as many as six impossible things before breakfast.

—Lewis Carroll

"**A** square's a square. You're all alike," the isosceles triangle shouted.

" Stop rubbing it in. I know we're each similar to one another," square replied.

"Similar? That's an understatement. All squares are alike if you know what I mean. You're...you're so conforming, so blah, just plain dull," rectangle joined in. "Your sides are always congruent to each other. Your angles are all always 90°. You're so predictable," rectangle teased.

"You always have four right angles too," square countered.

"Right, but at least my adjacent sides come in various lengths, and not every rectangle is similar to every other," rectangle fired back.

Circle suddenly barged onto the scene. "Look at me, square. I'm gorgeous—so round and smooth."

"But all circles are similar to each other," square interrupted.

"As I was saying," circle continued, "I'm so perfectly round, yet hidden within me is one of the most famous numbers ever discovered."

"Oh, we've all heard all about π from you before," square sighed.

"Yes, but we circles define π. If it were not for us, there would be no π," circle interjected.

Square got a funny look on its sides, and an idea flashed across its surface.

"All squares are ... so conforming, so blah, just plain dull."

"What is it?" circle asked. "Are you all right?"

"I'm more than all right. I'm incredible, amazing, and just as special as you are circle," square shouted joyfully.

"Come off it. You're the same old square," rectangle taunted.

"I agree. I am that, but we squares have within our boundaries a famous number. In fact, I'm responsible for the discovery of the first number of its kind," square said happily.

"Go on! What are you talking about?" triangle asked.

"I'm talking about the first number to be proven irrational. The $\sqrt{2}$. The square root of 2, the first number of its kind. It so happens to be found in me—in my diagonal's length. Every square's diagonal happens to be the length of its side times √2. Every square's diagonal divides the square into two identical triangles, whose right angles make each fit the bill for the Pythagorean theorem, which computes the length of my diagonal using my sides length. Voilá! For me, $1^2 + 1^2 = $ diagonal^2. So diagonal $= √2$," square concluded.

"Just hold it right there," circle interrupted. "Prove your diagonal is irrational. Show it can't be written as a fraction." Circle was certain square would be defeated by this challenge.

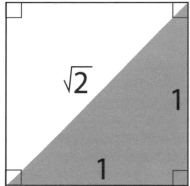

Square, without pausing, started in with a proof. "No problem. I'm sure I can prove this indirectly by assuming my diagonal's length is a rational number. If this leads to a contradiction, which I know it will, the proof will be done."

"What do you mean, a contradiction?" Circle was beginning to wish it hadn't challenged square.

Instead of belittling circle, square just said, "A contradiction contradicts a mathematically known fact. If √2 is rational, it can be expressed as a fraction in lowest terms. Call it p/q where p and q are whole numbers with q≠0. Then..."

"I believe you! I believe you!" circle shouted. "I concede your diagonal is an irrational length. And I take back you are just a square."

"You better believe that," square said with more confidence and pride than ever before. "Every square is not just a square. It's also a rectangle, a rhombus, a trapezoid, a parallelogram. In fact, a square is a convex quadrilateral." Square ended by shouting, "We squares cover all bases".

ANNOTATION_____

Square's indirect PROOF that √2 is an irrational number.

Suppose √2 is a rational number, namely p/q (in lowest terms), and if this proof leads to a contradiction, then √2 cannot be irrational.

p/q represents a rational number in lowest terms means p and q are integers with no common factors and $q \neq 0$. So we call this √2.

1) $\sqrt{2} = p/q$

2) $2 = p^2/q^2$, by squaring both sides.

3) $2q^2 = p^2$, by multiplying both sides by q^2.

4) $2q^2$ is an even number since 2 is a factor.

5) This means p^2 is even since it equals $2q^2$.

6) p must also be even since its square is even.
   This means 2 is a factor of p.

7) So p=2n where n is a integer.

8) Squaring both sides $p^2=4n^2$, now we substitute p^2 with $2q^2$ using the information from step 3. We get $2q^2=4n^2$.

9) So $q^2=2n^2$, by dividing both sides of step(8) by 2.

10) This means q is even since its square is even.

11) Steps (6) and (10) mean both q and p are even. So 2 is their common factor.

12) But wait, they can't have any common factors because they were in lowest terms to begin with. Here's the contradiction.
QED: So p/q which equals √2 can't be rational, so it must be irrational.

Fuzzy 1 clears things up

I have discovered such wonderful things that I was amazed…out of nothing I have created a strange new universe.

—Janos Bolyai

"**1** am having a really bad hair day. Any which way I turn and see myself in the mirror I look like a weird and blurry number," the pathetic looking number said in a wistful voice. "I feel so confused. I can't make out which number I am supposed to be," it continued.

"You're a mess!" 5 shouted when it saw the number's reflection in the mirror. "Just look at you! You're strange! In contrast, I am 5. I am plain, clear and simple. Everyday no matter what, I am 5 regardless of the time I get up or which side of my bed I roll out of. Regardless of the problem I appear in, I remain the amount 5 exactly and clearly. In spite of what is going on around me, I don't change. I am confident I will always be 5. But you are confusing. One moment you appear as 1 and the next as 0.1, and then 1.6. I can't figure you out."

At that moment the number 1 walked over to join this loud discussion. Turning to both the unknown number and 5, it replied, "What's all the fuss, you two?"

5 turned to 1 saying "You appear as clear and distinct as I do. But look at this weird looking number." 5 pointed to the pathetic unknown number.

"My Lord! As you see I am 1, and 1 is always 1." Pointing to 2, 1 continued its train of thought, "Look at 2, you are always 2. You 5 are nothing but 5. 1/2 is 1/2. A googol is a googol. $\sqrt{7}$ is $\sqrt{7}$. None of us is ever any other amount than the number we are! That's that." 1 said as it pounded its hand on a nearby table.

The unknown number's head began throbbing. It felt a terrible headache coming on, but even so it had to speak up. It had to make itself heard. It knew its self esteem depended on its explanation. Unfortunately, it could only muster up a very soft and timid voice as it said, "Are you sure?"

"Sure of what? Who said that?" 1 asked.

"I did," unknown number said turning to face 1.

"Where did you go? I can't see you," 1 continued.

"I'm right here in front of you," unknown number said moving forward, but becoming even fuzzier from its building headache.

"Is there something wrong with my eyesight?" 1 wondered aloud.

"No, I'm usually a little ambiguous," unknown number replied.

"A little? You are downright baffling. You can't be a number?" 1 said.

"Yes I am! But for some reason I am not necessarily specific." unknown number added.

"Not specific? All numbers are specific. What's your value?" 1 demanded. "You're not a variable are you?" 1 asked in the same breath.

"No, I'm not a variable. But I am not a specific number. I can be a whole range of numbers depending on the problem I appear in." With this, unknown number sounded even more confusing.

"...fuzzy numbers are a whole new species of numbers."

"That's impossible!" 1 shouted. "Every number has only one value, one amount it can represent."

"Well, that's true for you types of numbers," unknown number said. "But, we fuzzy numbers are a whole new species of numbers."

1 and all the other numbers stood slack jawed and speechless.

"That's my problem. Somedays I'm not certain which numbers I represent. They call me fuzzy 1." unknown number continued.

"Fuzzy 1 is not a name for a number. You sound like the rhyme 'fuzzy wasn't fuzzy was he'," 1 said making all the other numbers laugh.

"Well, you may be laughing at me, but you are all just ignorant," fuzzy 1 said with a forceful tone to its voice.

"We're ignorant?" 1 asked. "Why you just said you didn't know what number you represented. Right? Now who's ignorant."

"Actually I'm a number that came out of the blue?" fuzzy 1 explained.

"Out of the blue. What are you, some magic trick?" 1 teased.

"Like you, I represent the concept of an amount," fuzzy 1 answered.

"A very unclear concept, if you ask me," 1 replied. "You're not a specific amount, so how can you be a number?"

"That's exactly the point. Fuzzy numbers cover a lot of territory," Fuzzy 1 said feeling better, now that the truth was out in the open. In fact, its headache began to clear as it explained what it was all about. "In certain problems, many amounts or values can end up being me, fuzzy 1."

"You are really confusing me," 1 said scratching the top of its numeral, and beginning to wish it had never entered in this discussion.

"Look, 1, I can understand how you are confused, but you go back thousands of years, when the first scratch in the dirt was drawn to represent you. You came out of the blue. You represented one single item. I on the other hand just go back to 1920, when multivalued* logic was first introduced in mathematics, and fuzzy numbers were eventually created. Look, a fuzzy 1 set is not just one item. In fact it depends on who is using and defining it for a particular problem. If rounding-off is used in a problem, such as finding the dollar amount closest to $1.25. That would be a fuzzy 1. 58¢ is also represented by a fuzzy 1. In fact all amounts from 50¢ to $1.49 are fuzzy 1s in this type of problem," fuzzy 1 explained.

"So for this type of problem a fuzzy 5 could be represented by any amount from $4.50 to $5.49?" 1 asked.

"Exactly! Another example may be parking time and payment. For example a fuzzy $1 may cover any time from 0 minutes to 45 minutes while fuzzy $2 could be for 46 minutes to 2 hours, and so on. For a chemistry problem involving milligrams weights different intervals of weight can to be used and represented by fuzzy numbers," fuzzy 1 added.

"Fuzzy numbers are really FUZZZZZZY!" 5 declared. "Now I see why you are unclear, or should I say so blurry."

"Ambiguous, " said 1.

"Strange," said 2.

"Just different, and VERY USEFUL," fuzzy 1 said proudly.

Fuzzy logic began in 1920 with logician Jan Lakasiewicz, who revised traditional yes-no logic to *"multivalent"* (multivalued) logic. In other words, yes-no logic did not quantify the degree of truth in a statement Multivalent logic was devised to measure different degrees of truth or falsehood. Fuzzy logic is also referred to as the logic of uncertainty. For example, if a man says "he is young" and no definite age is given, his 'youth' is relative depending on other factors, such as the age of the person observing him. In 1965 mathematician Lotfi Zadeh applied multivalued logic to set theory and developed the concept of fuzzy sets.

A *fuzzy set* is a group of objects in which the elements of the set are not exactly cut and dry. The set {1,2,3,4,5} is not a fuzzy set because it is composed of just the five elements shown. Another example of a non-fuzzy set is the set of rational numbers. Its definition is clear cut. This set includes only numbers that can be represented as fractions whose numerators and denominators are integers with the denominator never 0. On the other hand, a set described as the set of of young people is a fuzzy set. Since the age of the people of this set is not specified, the set's elements differ depending on whom is being asked who they would include in this set. Elements composing fuzzy sets are not clearly defined.

Today, fuzzy sets and fuzzy numbers are invaluable in the design of robotics and smart machines. These concepts are used in the design of digital cameras, in automotive cruise control, antilock braking and warning systems and airbags, and even in our washing machines and airconditioners.

Some knots
are not

*Mathematicians do not study
objects, but relations between
objects.* **—Henri Poincaré**

"**W**hat do you mean I am not a knot?" Curly asked Square Knot.

"Just that. I am a knot and you are not," Square Knot replied with a superior note to its voice.

"I look as knotty as you. No one has ever suggested I was not a knot. How dare you?" Curly replied.

"It is not a suggestion," Square Knot replied. "It's a fact. You just wouldn't hold up if I pulled your ends. You are an imposter. But you are not fooling me anymore. Not now that I have been learning all about my heritage." Square Knot continued to break the news to Curly. "What do knots do?" Square Knot asked Curly.

"Why, they hold things securely. They are used to tie shoe laces. They rig boats. Tie packages. Make bows. All sorts of fun things," Curly replied, imagining a knotty life of knot tieing.

"Correct. And how many of these things have you done so far, Curly?" Square Knot asked.

"None, because I am still too young," Curly slowly responded.

"No! Not because you are too young, because you are not a knot. So forget it. There is no way you can be used to tie packages, a gift, or moor a boat," Square Knot said in a mean way.

"Well, how can you prove I am not a knot," Curly insisted.

Curley

"Simple. I just pull your ends… like this." At which Square Knot grabbed Curly's ends and pulled. "…and you disappear."

Sure enough Curly was no longer a maze of ins and outs of rope but simply a straight rope. But as soon as Square Knot let go of Curly's ends, Curly bounced back like a spring to its old shape, since this was the only shape it knew.

"Now pull my ends," Square Knot insisted.

Curly pulled and pulled until Square Knot let out a yell. "Stoppp! You're squeezing me too hard. I don't disappear. I hold together. See! I am a knot," Square Knot said proudly.

Curly walked away, with a downcast look to its shape, thinking… "All these years, I thought I was a knot. No one ever said otherwise. What am I?" Curly wondered, deeply saddened by its identity crisis. He was walking ever so slowly when a voice broke through its thoughts.

"Hi! there. You're an awfully cute knot," the voice said.

"But I'm not a knot," Curly had to confess.

"You sure look knotty to me," the little girl replied. "I would use you on a gift any day. You're not like those ordinary square knots," the girl complimented Curly.

"But I tell you I am not a knot. Pull my ends. Go ahead. You'll see." Curly held out its ends to the little girl.

As she pulled them he disappeared before her eyes. "Wow! Cool!" she exclaimed. And when she let go, Curly bounced back to its old form. "You're a magician's knot!" the girl shouted in amazement.

> # *"But I tell you I am not a knot. Pull my ends. Go ahead. You'll see."*

"A magician's knot?" Curly asked.

"Yes! Magicians turn and twist ropes into forms like you for their magic tricks. I would love to keep you for myself," the little girl said. "And technically speaking, for your information, a square knot is not a *mathematical knot*," she added.

" I can't join you, but thank you for the wonderful news!   I need to square things with Square Knot, who told me I was not a knot." Curly went off to find Square Knot. But it first decided to do some research, so it googled *knots* on its iPhone, and did it get an eyeful of facts and information! The little girl was right. Square Knot was not a mathematical knot! Curly was now ready for its stand off with Square Knot.

"Oh, it's you again. I thought I finished explaining to you why you're not a knot," Square Knot said impatiently when it saw Curly.

"Yes, I certainly understand what you said, but you neglected to mention that I was a *magic knot, and that is very important to my self-esteem,*" Curly replied.

"Magic or not, you're still not a knot," Square knot said.

"Well, I have news for you, Square Knot. Neither one of us is a knot," Curly declared.

"What are you talking about?" Square Knot asked, certain Curly was at loose ends. "I'm a knot because I secure things and you are not a knot because you don't."

"As I was saying, before you rudely interrupted me—We are not mathematical knots because math knots don't have ends." Curly explained.

"I've never heard of a knot without ends," Square Knot replied emphatically.

"Well, that shows what you don't know," Curly countered. "Have you heard of a trefoil knot?"

"No," Square Knot replied.

trefoil knot

"Well, it's the simplest mathematical knot. It has three crossings. See here's how it looks. And it comes in left and right handed versions which happen to be mirror images of one another. "

"How can such a knot hold anything together," Square Knot asked.

"That is not a math knot's purpose," Curly said. "Mathematicians have shown that math knots cannot exist in more than three dimensions and many other things. They know there is only one math knot with four crossings, and two types of knots with five crossings, and thousands of knots with thirteen or less crossings. "

"Hold it right there. What is the purpose of these math knots since they can't tie things together?" Square Knot challenged Curly.

"They don't tie physical things together such as shoe laces and boat moorings" Curly began to explain, "but they tie concepts together. Today knots are used in the study of molecular biology and physics. These knots are linked to the study of

DNA configurations, so that makes them invaluable in genetic engineering. In physics, their shapes can resemble the interaction of particles. Mathematicians and computer scientists are uncovering and classifying knot properties that do not change as they go through various transformations. Mathematical knots are also being reduced to math expressions such as the Alexander polynomials."

"Stop, stop, stop!," Square Knot yelled. "You're making me dizzy and confused with all this stuff on math knots. I agree we are not knots in the world of mathematics."

"Good," said Curly. "So you should also be informed that I am not only called a magician's knot, but a Chefalo knot and a false knot. I am so interesting that I have three knot names," Curly said enjoying rubbing it in to Square Knot.

For once Square Knot was at a loss for words.

Curly took off saying, "Ta ta for now. I am going off to join a magic act in a circus. I'll be really busy even though I'm not a knot. See you around." Curly, the magic knot, walked upright and proudly… having just discovered itself.

The dilemma of the order of operations

"Mathematicians are like Frenchmen: whatever you say to them they translate into their own language and forthwith it is something entirely different.
—Johann Wolfgang von Goethe

A long time ago when numbers and operations, such as +, −, x and ÷, were just beginning to mingle, something very strange happened one day. But let's backtrack. Until this day, people had been using +, −, x and ÷ with numbers in the following types of problems— 6+2 or 6x2 or 6-2 or 6÷2. There was no question of what had to be done to solve these problems.

On that day the lazy mathematician Nonchalant did not want to write out all the steps separately to solve a problem. Instead, Nonchalant just strung all his numbers and operations for his problem in one line, like this — 3+3x3.

You can imagine what the other mathematicians said. Precision exclaimed, "Nonchalant, that problem makes no sense! If I do 3+3 and then *times* this result by 3, I get 18 as my answer. But if I do 3x3 first then add 3 to the result, I get 12. The way you wrote this problem, how can I tell which answer is correct?" Precision was very perplexed.

"I know! Let's ask Decisive what to do," Nonchalant replied.

But when they brought the problem to Decisive, Decisive was dumbfounded. "The operations have been done correctly in both cases. Either answer is correct depending on how you

approach the problem. This is a *dilemma,"* Decisive concluded.

"Yet we can't let both answers be right. Mathematics needs to be more precise," Precision said. "We have no choice but to ask Arbitrary which to choose as the correct answer." So off they went to Arbitrary for advise.

Arbitrary thought, and thought, and thought about the problem. Finally, his eyes lit up, and he said, "I'll just have to arbitrarily make up a rule that we must all agree to follow whenever many operations and numbers are joined together." After declaring this, Arbitrary said, *"Whenever you have numbers and operations in a problem we must agree to first do multiplication and division before addition and subtraction always working from left to right.* How does that sound?"

Precision said, "That sounds fine." Decisive also agreed. Nonchalant said, " It doesn't make any difference to me which rule we decide to follow."

"Wait!" shouted Parentheses. "The rule has a flaw."

"A flaw?" the rest asked, startled.

"Yes, we need to add one thing," Parentheses said.

"What is it?" the others inquired anxiously, waiting to hear what Parentheses had to say.

"I would restate the rule as follows: *Whenever you have numbers and operations in a problem we must agree to first do the operations that appear in parentheses following the rule of doing multiplication and division before addition and subtraction and always working from left to right.*

"That's it!" declared Arbitrary.

"Wonderful!" said Precision.

" Good idea," said Decisive.

"Makes no difference to me," replied Nonchalant.

So the dilemma of *the order of operations* was solved.

Transfinites make their entrance

Counting is a less precise tool for infinite sets than for finite ones. The shepherdess who can count her flock of a hundred sheep will know if the wolf has taken one; but, if she has an infinite flock, she won't notice until almost all of her sheep have been lost.

—Peter J. Cameron

$\aleph_0$

$\aleph_1$

$\aleph_2$

$\aleph_3$

$\aleph_4 \cdots$

This particular day the counting numbers were hanging out together in their set.

"What do you think those are?" 7 wondered as it glanced to the right. "I've never seen anything like them. Do you know what they are?" 7 asked.

"I haven't the faintest idea'" 1 replied.

"They sure look ominous." 7 added.

You could tell 3 felt frightened as it moved and pulled 2 and 4 closer to it.

The strange objects began moving toward the counting number set, at which the counting numbers inched backwards.

"Stop!" 2 yelled. "Don't come any closer. Who or what are you?"

"We're special numbers." $\aleph_0$ replied.

"Numbers?" 1 said, slack jawed. "I have been around for thousands of years and I have never seen any numbers resembling you."

"We are the transfinite numbers." $\aleph_0$ said.

"Trans-who?" 1 asked.

" The transfinite. Our name describes us perfectly." $\aleph_0$ continued. "We describe or should I say count or assign a size to infinite sets, such as your set."

"Come on," 2 said. "You're joking. Who came up with that great idea?" 2 asked sarcastically.

"Why, Cantor did," $\aleph_0$ said. "George Cantor. He invented or should I say discovered us to describe and measure such infinite sets as the counting numbers, the whole numbers, even the integers."

"Well, which of you size up the counting numbers set, and which the whole numbers set? " 1 asked.

"Why, I do. I am called aleph-null," $\aleph_0$ replied.

"You mean you are used for both sets? Clearly the whole numbers set has one more element than the counting numbers set, namely zero. Since the whole numbers have the zero, they have one more element and should be counted as larger. So how can you be used to measure both sets?" 1 asked.

"Well, the brilliant Cantor explains why I am used," $\aleph_0$ replied. "See how the counting and whole numbers numbers are listed in ascending order," $\aleph_0$ said while whipping out its chalk and black board and wrote the following:

{1, 2, 3, 4, ...} *the counting numbers*

{0, 1, 2, 3, 4, ...} *the whole numbers*

"Yes, that is correct," 2 replied. "So?"

"So Cantor proved both sets have the same number of elements even though it appears the whole numbers have an additional number, the number zero." $\aleph_0$ tried to explain.

"How?" demanded 1, who was beginning to get a little impatient.

"Cantor made a one-to-one correspondence between each element of these sets, like this:

{1, 2, 3, 4, 5, ...} *the counting numbers*
  | | | | |
{0, 1, 2, 3, 4, ...} *the whole number*

Each element from the counting numbers is matched with only one distinct element, which always happens to be one less than the element of the counting numbers. It works perfectly. Since both sets are infinite, neither set will run out of an element to match up with the other. Another way to see this is to show that any counting number, call it **k**, will match up with the whole number **k-1**. And so, the matching process never fails, since the sets are infinite. Therefore both sets have the same number of elements, and this number is referred to as the cardinality of the set. I am the transfinite number that describes this infinite cardinal number." $\aleph_0$ explained.

"We are amazed," the counting numbers said in unison.

"Cantor was a genius," 2 said. "So which of you transfinites describes the number of elements for even numbers?"

"I still do," $\aleph_0$ replied.

"What? How can that be? The evens should have a lot less, since there are no odd numbers present. In fact, I think it should have half the number of elements of either the counting or whole number sets," 2 declared, beginning to feel confused.

"Here, let me explain. Look at this one-to-one correspondence," $\aleph_0$ pointed out on the black board:

$$\{1, \quad 2, \quad 3, \quad 4, \quad 5, \quad \ldots \quad k, \ldots\} \text{ the counting numbers}$$
$$\{2, \quad 4, \quad 6, \quad 8, \quad 10, \quad \ldots \quad 2k, \ldots\} \text{ the even numbers}$$

Each element matches up with one and only one from the other set."

"Wow!" 2 exclaimed. "So Cantor probably did the same sort of thing for the odd numbers."

"So the number of elements for the odd numbers are also described by $\aleph_0$," 3 said, feeling oddly satisfied with itself.

"Exactly" $\aleph_0$ said while writing out the following one-to-one correspondences:

$\{1, \quad 2, \quad 3, \quad 4, \quad 5, \quad \ldots \quad k, \ldots\}$ *the counting numbers*

$\{0, \quad 1, \quad 2, \quad 3, \quad 4, \quad \ldots \quad k\text{-}1, \ldots\}$ *the whole numbers*

$\{1, \quad 3, \quad 5, \quad 7, \quad 9, \quad \ldots 2k\text{-}1, \ldots \}$ *the odd numbers*

$\{2, \quad 4, \quad 6, \quad 8, \quad 10, \quad \ldots 2k, \ldots \}$ *the even numbers*

"Since we know the counting numbers, the whole numbers, the evens are all counted by me, then any set that can be put into a one-to-one correspondence with any of these sets will have $\aleph_0$ number of elements," $\aleph_0$ explained. "Any such set is called a *countable* set."

" We now see how the odd numbers have $\aleph_0$ elements," the odd numbers replied.

"What? How can that be? The evens should have a lot less, since there are no odd numbers present. "

"We are astonished!" each of the four sets said. "But what about the integers. Surely, they don't have $\aleph_0$ number of elements."

"It's also $\aleph_0$," $\aleph_0$ said proudly.

"I can't see how that's possible." 1 scratched its head. "How can a one-to-one correspondence be shown between the set of integers and a set that has $\aleph_0$ number of elements?"

"Just let me rearrange the whole numbers in this manner, and you will immediately see how this is possible."

So $\aleph_0$ wrote out the even numbers from the set of whole numbers numbers on the left side of zero in descending order and on the right side of zero its odd number in ascending order. Like this:

$$\{...,-2k, ...\ 8,\ \ 6,\ \ 4,\ 2,\ 0, 1, 3, 5, 7, ..., 2k\text{-}1, ...\}$$

And then it placed the set of integers above this rearranged set of whole numbers:

$$(...,-k, ... , -4, -3, -2, -1,\ 0,\ 1, 2,\ 3,\ 4,\ ..., k, ...\}$$
$$\{...,-2k, ...\ 8,\ \ 6,\ \ 4,\ 2,\ 0,\ 1, 3,\ 5,\ 7, ..., 2k\text{ } 1, ...\}$$

"Incredible!" 1 shouted. "It's incredibly clear. I can easily see how the whole numbers and the integers match up. That's an excellent explanation," 1 added, patting $\aleph_0$ on its shoulder.

"I can't take credit for this explanation, but it is a beautiful way to show it," $\aleph_0$ replied with a big smile. "Cantor even came up with a wonderful proof of why the fractions, aka rational numbers and even algebraic numbers have $\aleph_0$ number of elements."

"You don't have to get into all those sets now, I've understood what you said so far, so don't push it," 1 said. "But I am still confused about something else," 1 said. "How are transfinites used? Why in the world did Cantor want to discover such ideas?"

"Because we were concepts waiting to be discovered in the *world of mathematics*," $\aleph_0$ began explaining. "The problem of infinite sets were a huge challenge Cantor created for himself," $\aleph_0$ said. "The usefulness of transfinite numbers never entered into Cantor's decision to work with infinite sets. Mathematicians' goal is to solve problems, and the more difficult the better. Why do you think they spend decades and even centuries trying to solve some problems…it's not necessarily the answer per se, but the process and what is learned as they pursue the answer."

A hush came over the counting numbers. They did not know what to say or what to think.

"I know people came up with counting numbers because they needed us to count things," 2 finally spoke up. "We immediately had an important role to play in their everyday lives. We keep track of amounts for them. And then as they ran into problems where we could not furnish the answer, they invented new numbers such as fractions and irrational numbers. Surely transfinites have some role."

"Look," $\aleph_0$ began. "We may or may not have a role some day in the real world, but we do exist in the world of mathematics, and that is what is important to us."

"I know," 2 interrupted. "Let's pursue that some other time. But I still have a question," 2 said. " What do the transfinite numbers $\aleph_1$, $\aleph_2$, $\aleph_3$, $\aleph_4$, ... and all the rest count?"

" I knew one of you would ask that," $\aleph_0$ said. "Let it suffice to say Cantor discovered a lot of stuff about transfinites. He showed that the number of elements[1] for real numbers are not countable, meaning they cannot be placed in a one-to-one correspondence with any set," $\aleph_0$, explained. $\aleph_1$ is used for real numbers, and $\aleph_1$ also describes the number of elements that are points on a line, on a plane, and many other things. He created an entire arithmetic for us transfinite numbers, but that's a whole other story."

"Let's talk about that another day," 2 replied feeling its mind had reached its saturation point. So the transfinites slowly exited the area.

---

[1]Cantor defined that two infinite sets have the same transfinite number If their elements can be put into a one-to-one correspondence with each other. The cardinal number describing both of these sets is the same number, namely a transfinite number. An infinite set is *countable* if it's transfinite number is $\aleph_0$. Cantor developed an ingenious proof involving what is called a *diagonal argument (see page 116)* to illustrate how the rational numbers could be put into a one-to-one correspondence with the counting numbers, and, thereby, illustrating these two sets had the same transfinite number, $\aleph_0$, of elements. Using an indirect proof Cantor also proved the number of elements in the set of real numbers is a greater transfinite number than $\aleph_0$. Cantor showed that the real numbers which were the union of the rational and irrational numbers could also be described as the union of the algebraic and transcendental numbers. He proved that the algebraic set's transfinite number was $\aleph_0$, making that of the set of transcendentals a larger transfinite number. In addition, he also explained the existence of infinitely many transfinite numbers.

Numbers, the number line, points & all that stuff

*What? Will the line stretch
out to the crack of doom?*

—William Shakespeare

Macbeth

In the beginning line was just an infinite set of points. No fancy frills. Just two arrows on either side indicating how it goes on endlessly and perfectly straight in opposite directions. Line was happy with just its points to keep in line. But slowly things began popping up above and below its points. Mathematicians were placing things on them. Things they called numbers. It was strange having 1, 2, 3, 4, … all equally spaced at points along line. To line they felt like intruders. But mathematicians were not satisfied with just these numbers so they added 0 and the rest of the integers—all equally spaced.

Not much later, fractions appeared between the whole numbers and the integers. Fractions started out acting timid and shy around them. In fact, they felt incomplete and outnumbered by the integers. But, when mathematicians began to average two or more integers new fractions were often created. In fact, mathematicians found that between any two numbers on the number line there was always another they could find by averaging. Sometimes the averaging resulted in a fraction and other times it was an integer. But the number of fractions used were definitely increasing. Below every number was a point that went along with it, and the absolute value of the point's number measured the point's distance from the origin. The integers began getting a bit concerned. And concerned they should have been, since fractions appeared to be taking over line. Jamming themselves between all the integers, the integers were feeling more and more uncomfortable. Their anxiety came to a head when integers realized that between any two integers an *infinite number* of fractions existed. This fact freaked out the integers. Needless to say the integers felt they were outnumbered. Fractions on the other hand began to feel empowered by their sheer numbers, even though fractions such as 1/2, 1/3, 1/4, 1/5,…, 2 1/2,… had not yet realized there were infinitely many fractions between each pair of consecutive integers.

Imagine fractions pushing integers around—this was unheard in the number world. Soon, integers felt intimidated whenever fractions began clustering around each of them and settling on more and more points around them. Line felt an overpopulation crisis in the process.

Little did integers or the fractions suspect what was in the offing. Mathematician Georg Cantor was busy at work. Work that would actually end up equalizing the world of integers and fractions. He brilliantly figured out a way, indeed he actually proved, that there is the same quantity of integers and fractions. But how could there be a number to describe this quantity since there were infinitely many of each?[1]

1 was the first integer to break the news to the other numbers. "Impossible!" 1/2 shouted at 1 on its left on line.

"It's all here in Cantor's proof," 1 replied quietly but firmly even though 1 did not understand Cantor's proof.

"Let me see that," 1/2 said as it grabbed the paper from 1. "Countability, countable sets, one-to-one correspondence — gibberish," 1/2 said as it threw the paper down.

"Just a minute!" 2 exclaimed. It then reached down to pick up the proof, and addressed 1/2. "I have experience with mathematical proofs. This is right up my line. I've sure appeared in proofs with primes to even numbers to the irrationality of the square root of 2." Studying the paper 2 finally said, "This mathematician's proof may appear confusing, but in the end it sounds logical." 2 then proceeded to explain Cantor's proof *.

"Cantor takes us integers and matches each one of us up in an ingenious way with just one of each of the fractions. He left none of us or them out. Every integer has just one fraction associated with it and vice versa. It's brilliant!"

At the end of the explanation when 1 heard the QED, it stood a little straighter and taller. 1/2 and the other fractions, on the other hand, looked dejected realizing fractions had lost their edge over the integers.

0, which was neither negative nor positive, tried to strike an upbeat note by saying, "We should all be happy with this result. We're balanced. We're equal in number even though that number has an unusual name and symbol ...*aleph null,*

$\aleph_0$. "It's a balance of power," 0 explained. "No one feels intimidated or outnumbered anymore. We've found harmony. Now let's all find our places on the number line."

As the integers and fractions found their respective points on line, it became apparent that not all points had number

"Dense? How dare you," 2 bristled.

"No! no! no! Not thick headed…" line answered.

names. "Wait just a minute," a voice hidden in the points spoke up.

"Who said that," 1/2 asked. "None of us," the integers and other fractions replied.

"It's me and all of us," the voices of the unnamed points chimed in.

1 and 1/2 and 0 looked at one another asking, "Do you see anything?"

"Just points," 2 replied.

" Exactly!" line replied. "Numbers other than integers and fractions name the rest of my points. You think you numbers are the only numbers? Well, I have news for you. You are dense."

"Dense? How dare you," 2 bristled.

"No! no! no! Not thick headed, though at times you sure sound that way," line answered. "Dense means that between any two of you there is always at least another rational number."

"Rational?" 0 asked.

"Yes, what you all are is *rational*, and I don't mean *sane*. In math it means that every one of you can be written as a fraction. For example, 1 is equal to 1/1, or 5/5, etc. 8 is equal to 8/1 or 16/2 etc." line explained. "That's what rational means."

"What about me?" asked 0.

"One of your fraction equivalents is 0/1," line replied. "Every one of you can be expressed as a fraction, which mathematicians also refer to as a repeating decimal."

"But how does that help us answer what the names of all the remaining blank points are," 1 asked.

"I'm surprised you would ask," line said. "You, of all numbers, were the first. You should know the history of numbers."

"I'm very old and my mind is not what it used to be," 1 answered line.

"Come off it," line replied. "No, no not off the line, I meant hogwash. Your memory is not failing you. You just didn't pay attention when mathematicians discovered the irrationals which make up the rest of the reals. These include $\sqrt{2}$, π, e, $5\sqrt{2}$, phi, et cetera[2]. As you can see by the remaining unnamed points there are as many of them as there are you. Actually there are more, which has also been proven by Cantor."

The rationals were amazed. They had no idea they were dense and that there were more of these other numbers than there were rationals.

The whole numbers still didn't get what line was saying. "We must also be dense," the whole numbers said in unison.

"On the contrary, whole numbers are not dense as whole numbers, but only when they're considered rationals," line said trying to help the whole numbers understand. "For example, if you take the whole numbers 3 and 4 there is no other whole number between them, but on the other hand when you are thought of as rational numbers, there are other rationals between 3 and 4," line explained patiently.

"What makes you so knowledgeable?" 1/2 asked line.

"I am a continuum. I am c," line declared.

The rationals looked at one another and felt line had lost it. "What are you talking about," 2 got up the courage to ask.

"Remember George Cantor? Well he also proved that the number of real numbers is the same as the number of points on me. I am in fact the real number line," line proudly declared. "And I am called the continuum, designated by the letter[3] c. Think of me as the home of the reals which include all rationals, irrationals and transcendentals."

All the numbers looked at one and other with expressions of awe sweeping across their symbols. Finally 1 broke the silence by addressing all the numbers and line. "I guess we all learned something along this line today." At which a roar of laughter broke out, which immediately eased the tension, giving line the opportunity to yell, "So let's all take our places. Line up and look sharp."

---

[1] See *The Transfinite* on pages 99-107

[2] Some irrational numbers are transcendental numbers. Transcendental numbers are not roots (or solutions) of a algebraic equations with rational coefficients. An algebraic equation is of the form $a_0x^n + a_1x^{(n-1)} + a_2x^{(n-2)} + ... + a_n = 0$, where $a_0, a_1, a_2, ...$ an are rational numbers. $\sqrt{2}$ is irrational, but not transcendental because it is a root of $x^2 - 2 = 0$, while π and e are irrational numbers which are also transcendental because there is no algebraic equation (aka polynomial) with rational coefficients that has π or e as roots.

[3] Continuum c is not to be confused with **c**, the speed of light in a vacuum from Einstein's $E = mc^2$.

*ANNOTATION
Cantor devised a brilliant method to show that the cardinality for the rational numbers, which are all numbers that can be expressed as fractions which include integers, is the same as that of the integers. Both the rational numbers and the integers can be put into a one-to-one correspondence with the natural numbers. To do this Cantor invented a method for arranging the rational numbers in a list so that every rational number would appear somewhere in this arrangement. The chart below shows Cantor's method for listing all rational numbers.. The dotted line shows his method for selecting each and every rational number. The one-to-one correspondence below the list shows how he matched up each natural number with exactly one rational number.

$$1/1 \quad 1/2 \quad 1/3 \quad 1/4 \quad 1/5 \ldots$$
$$2/1 \quad 2/2 \quad 2/3 \quad 2/4 \quad 2/5 \ldots$$
$$3/1 \quad 3/2 \quad 3/3 \quad 3/4 \quad 3/5 \ldots$$
$$4/1 \quad 4/2 \quad 4/3 \quad 4/4 \quad 4/5 \ldots$$
$$5/1 \quad 5/2 \quad 5/3 \quad 5/4 \quad 5/5 \ldots$$

$$\{ 1, \quad 2, \quad 3, \quad 4, \quad 5, \quad 6, \quad 7, \quad 8, \quad 9, \ldots \}$$
$$\{1/1, \ 2/1, \ 1/2, \ 1/3, \ 2/2, \ 3/1, \ 4/1, \ 3/2, \ 2/3, \ldots \}$$

Cantor also devised an ingenious way to show that the irrational numbers outnumber the rational numbers. He came up with a proof using a methods referred to as *diagonalization*.

The mathematical taboo

I don't agree with mathematics; the sum total of zeros is a frightening figure. —Stanislaw J. Lec

More Unkempt Thoughts

"I'm zero! I'm zero! I've finally been recognized as a number!" zero shouted jubilantly.

"But you're nothing, zilch, nil, nada, " 1 taunted the exuberant zero.

"But I am a number. I've finally been discovered. I'm the number that makes the set of counting numbers the set of whole numbers," zero replied with assurance. "I'm the origin on the number line."

"But you're worth nothing," 1 continued to tease.

Zero was too elated to let 1 continue. "I'm more than -1 or -2 or any negative number. Right? So I'm worth something. If a budget equals 0, it's balanced. Right? I'm the dividing point on a thermometer with readings so many degrees above or below 0. I am important!" zero attested.

"Yeah, but what would someone rather end up with 0 dollars or 1 dollar?" 1 asked.

"Well, if they ended up with 0 dollars, at least they would have spent their money and enjoyed it," zero fired back.

But 1 wouldn't drop it. It wouldn't let zero have its moment of triumph. "You're still a pain in the neck to mathematicians because you're the perpetrator of the mathematical taboo."

"You would have to mention that, wouldn't you?" zero lamented.

"At least with me there's no problem doing any operations," 1 continued.

"Yes, but you're boring. Look, 1 times 10 is 10. 1 times 8 is 8. 10 divided by 1 is 10. 8 divided by 1 is 8. And 1 added or subtracted from any number is no big deal…the result is just more 1 more or 1 less. You're just plain BORING 1!" zero could no longer hold back its anger.

"It's a mathematical yellow traffic signal, not a taboo."

"How dare you call me boring….I am the first number ever used."

"Boring, boring, boring!" zero taunted. "Not that old argument again," zero continued. "Sure you're the first number people came up with, but you're definitely not the last. And there are numbers just as famous and more sophisticated than you."

"Like what?" 1 yelled.

" Look at pi, e and phi, or even i. You have to admit they are far more interesting numbers," zero replied.

"Such insults...to be compared to irrational, transcendental, and imaginary numbers. I may be boring, but I am not feared when it comes to division," 1 confronted zero.

"Okay, I concede. Mathematicians never like to divide by me. But they do love me in multiplication. The moment they see me they know the answer to their problem. For example, (3x2.9805)(476)(2/3)(0) they immediately know it's me."

"But what about the taboo," 1 shouted so loudly it shook the number line.

"Sure 1/0 is impossible* to complete, is undefined, is

---

* ANNOTATION

mathematical taboos

What is a mathematical taboo? It is a process which creates a false statement or impossible answer. For example, dividing by zero can create the false statement 1=0. Suppose we are allowed to divide 1 by 0, then 1/0 equals must equal some number, call it "a".
So, 1/0 = a,
which means 1=a•0, but then 1=0, since 0 times any number a, is 0.

There are other no-noes in mathematics. But dividing by zero is probably the best known "taboo". Another way to look at the problem of dividing by 0 is to consider the problem *6 divided by 3*. We know it takes two 3s to make six. But, if we consider dividing 6 by 0, how many zeros does it take to make six? No matter how many zeros you add together you will never get 6. This is also known as a *mathematical infinity*.

labeled as an infinity, but this result just tells mathematicians either they have to approach the problem in another way or they did something wrong. It's not a taboo. It's just a flag. A signal that something was wrong in their reasoning. Just like a traffic signal…if it's red, we stop; yellow, proceed with caution; green, go on. It's a mathematical yellow traffic signal, not a taboo."

"But…," 1 began again.

"But nothing," zero interrupted the frustrated 1. "I'm tired of our bickering. I am putting an end to this argument." And with that, zero multiplied itself by 1. The result, as we know, was 0.

1s and 0s take on the universe

I see a certain order in the universe and math is the one way of making it visible. —May Sarton

Have you heard the latest?" 7 asked.

"The latest about what?" 9 replied.

"About 1 and 0," 7 whispered.

"What are you whispering about? Sure I know 1 is greater than 0," 9 declared emphatically.

"No. No, silly… about the binary number system of 1s and 0s?" 7 said urgently, but still in a quiet voice.

"I've known about the binary system for ages. Everyone does. Get with it 7. That's the way numbers are written using just 1s and 0s, " 9 explained, feeling annoyed.

"No, no no!" 7 shouted this time. "They think they are everything."

"They have always believed they were hot stuff," 9 replied. "Since their binary system works perfectly with electricity, they think they govern everything —computers, DVDs, CDs, traffic signals, cameras. More and more things are getting digitized, and 1s and 0s are in charge."

"No. It's more than that," 7 continued. "Some mathematicians believe they are the universe."

"That's ridiculous," 9 said as it began walking away.

"Hold it! They say 1s and 0s make up everything we see, feel and sense in the universe. 1s and 0s make up the ultimate 'virtual' reality," 7 tried to explain.

"You mean there is no pain, no joy, no flesh and blood, no matter? It's all our imagination? Everything is just 1s and 0s...yeses and noes?" 9 gasped. "That is impossible!"

"Not according to some mathematicians. They feel the universe is a cosmic computer set in motion by a few simple rules, like cellular automata* one can watch run on a

"Everything is just 1s and 0s...?" 9 gasped. "That is impossible!"

"Everything is just 1s and 0s...?" 9 gasped. "That is impossible!"

computer monitor. You, me, people, dogs, cats, plants, matter and antimatter...everything is 0s and 1s." 7 was now shouting.

"I can't believe it," 9 said, shaking the top of its symbol.

"We'll you're not the only one. But it gets worse," 7 continued. "Some believe computers exist everywhere. The body is run by a computer—the brain. Mathematicians have now figured out how the DNA of a cell can be used as a computer.[1] They are making optical and even quantum computers. Even the atoms in a cup of coffee can be harvested into a computer. And they believe they are equivalent—all part of the cosmic computer."

"Where does the cosmic computer come from?" 9 wondered.

"As I said before, from a simple set of rules, similar to algorithms mathematicians use to create evolving cellular automata," 7 answered.

"You mean all living things and inanimate things and how they function are types of 3-dimensional pixels

---

*ANNOTATION_____

The idea of cellular automata originated with Konrad Zuse and Stanislaw Ulman in the 1940s. Then in the 1960s, Zuse and Ed Fledkin both independently conceptualized the idea that cellular automata were at the crux of the universe. In 1970, fuel was added to this theory when mathematician John Conway devised the *game of life*, a simple computer model using cellular automata that mimic the growth and evolution of living entities. In the 1980s, the work of Stephen Wolfram boosted interest in cellular automata showing how a set of simple rules could generate complicated and complex objects. In 2002, Wolfram published *A New Kind of Science* in which he views the operation of the universe as an enormous cellular automaton—something like a gigantic Turing machine. In 1937, Alan Turing and Alonso Church proved any computation done by a finite-state machine could be done by any other finite-state machine on an infinite tape. This explains why Mac's and PC's work can be interchanged once certain software has been installed. The Turing machine is a theoretical computing machine capable of doing any operation a human can and able to imitate the operations of any computing machines.

brought about by using 1 and 0 in a fundamental set of yes and no rules?" 9 asked 7.

"That's what they contend," 7 replied

"So what about all the other cosmic theories with atoms, quarks, quantum physics, vibrating multi-dimensional strings, dark matter, dark energy, …"

"The 1s and 0s," 7 interrupted, "are the ultimate essence of these. In the beginning, there was 1 and 0. They got together and made strings of 1s and 0s. These formed words which became rules. These rules became a cosmic program. The rest is history."

"I don't think I can live with them…1 and 0 will be insufferable," moaned 9 as it walked away shaking its head.

---

[1] In 1994, computer scientist Leonard M. Adleman of USC demonstrated how DNA could be used to form a molecular computer to solve a mathematics problem. For further information, see pp138-140 *Mathematical Footprints*, by Theoni Pappas, Wide World Publishing, San Carlos, CA, 1999.

How the 4th dimensional cube got its nickname

Mathematics may explore the 4th dimension and the world of what is possible, but the Czar can be overthrown only in the 3rd dimension.
—Vladimir Ilyich Lenin

One autumn day many years ago a cube was sitting on a log watching the leaves fall to the ground, and thinking, " I am so lucky to be 3-dimensional. I can look down on a plane and see points, lines, segments, triangles and all sorts of other flat objects without them even knowing I am looking in on them. I guess I am a kind of voyeur." Then it added, "I feel on top of the world."

"What do you mean? " a voice asked from out of nowhere?

"Where are you?" cube shouted, startled by the voice.

"I'm up here looking at you," the voice replied.

The cube looked up, and said, "Oh you're just a cube like me, but I am having trouble keeping you in focus. Why are you jumping around so much? You are very hyper. Can't you just sit still?"

"But I'm not a cube. You're just seeing some of my many impressions," the voice explained.

"What are you talking about?" the cube asked. "You look like parts of a cube constantly moving around and showing off your different parts."

But the voice said insistently, "As I said I am NOT a cube."

"Then what are you?" the cube demanded.

"Let me qualify my statement. I am not a cube like you, but a fourth dimensional cube. I come from a world totally alien to you. The fourth dimension!" the voice explained.

" Have you flipped? What do you mean the fourth dimension? I'm here in the third dimension, and I am looking at you jumping from one facet of a cube to another, as if you can't sit still. I think you're just a very hyper cube and having trouble remaining still. And what do you mean you're from another world. I am on top of the world. I see everything all around me and above and below me."

"Well, you don't seem to be able to see me as I really am," the voice replied.

"I don't understand. How come I cannot see you?" the cube asked.

"That's because you are not fourth dimensional, but merely a three dimensional object," the voice answered. "And I certainly am not hyper? I have never even been anxious. I saw you looking in on the points, lines, squares on a plane from your vantage point in the third dimension. I heard you thinking out loud how they did not even realize you were spying on them, and that is exactly what I was doing to you. Spying on you without you knowing. As you know any

2-dimensional object can only see part of you as you pass through its world of a plane. A square might see your corner as a point, your side as a segment, and your face as a square, but it would never be able to see you as a cube," the 4-D cube explained. "Do you understand ?" the voice asked the cube.

"I am beginning to get the picture," cube replied. "You are saying I see only parts of you."

"I am not a cube like you, but a fourth dimensional cube."

"Exactly! You may get a glimpse of a corner, a face, an edge, or one of the many cubes that compose me," the voice explained.

"Are many cubes composing you?" the cube questioned.

"Yes. I have 8 cubes," the 4-D cube replied.

"You are hard to imagine," cube said.

"Tell me about it. I am as difficult for you to imagine as a square is to a segment, or let alone a point that has no dimension. But some people have tried in many ways to come up with how my shape would look. So let me help out your imagination with these 3-D renditions of me," the 4-D cube said.

"Here is how I was pictured by architect Claude Bragdon[1] in 1913.

"And here is how I was mathematically unfolded into the 3rd dimension.

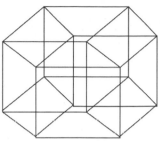

Bragdon's hypercube

Push me back into the 4th dimension and I am back to my old self.

the hypercube unfolded in the 3rd dimension

"In both these depictions you can pick out the eight cubes which compose me.

"With the advent of computers, scientists such as Thomas Branchoff captured slices of impressions of my parts as I was programmed to pass through the third dimension. Something like what you saw when you said I would not sit still and seemed to be constantly moving," the 4th-dimensional cube explained.

"Even though I cannot see you as you are in your dimension, I do feel I am beginning to understand. But I think I will always think of you as a hyper cube," the cube replied.

"Well, I do have another name other than 4th-dimensional cube. I am also referred to as a *tesseract*. So now I will also

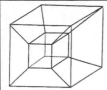

This rendition of a tesseract shows one cube in the center, six cubes coming off each of the center cube's faces, and one big cube encircling all these.

adopt your name "hypercube"[2] as one of my nicknames.

---

[1] Architect Claude Bragdon (1866-1913) was one of the first people to try to depict what a fourth dimensional cube would look like. This drawing is from his book *A Primer of Higher Space*.

[2] The term hypercube in mathematics refers to a n-dimensional cube. n represents any whole number. If n=1, this is a segment since it only has 1-dimension. If n=0, this is a point which is 0-dimensional. Since a cube is 3-dimensional, n=3. A hypercube has n=4 making it a 4-dimensional cube, and is called a tesseract. The number of vertices a hypercube has is given by 2^n. For a point, n=0, so $2^0=1$ the vertex. For n=1, $2^1=2$ which represent the two vertices of a segment. For a square n=2, so $2^2=4$, which are the 4 vertices of a square. A cube, n=3 has $2^3=8$ vertices, and a tesseract or a 4th-dimensional hypercube with n=4 has $2^4=16$ vertices.

•
a point is a 0-dimensional hypercube

•———➤
a segment is a 1-dimensional hypercube

hypercubes of different order and how the 4th-dimensional hypercube evolves from lower dimensions.

a square is a 2-dimensional hypercube

a cube is a 3-dimensional hypercube

a tesseract is a 4-dimensional hypercube

What a line!

...he seemed to approach the grave as an hyperbolic curve approaches a line, less directly as he got nearer, till it was doubtful if he would ever reach it at all.

—Thomas Hardy

Far from the Madding Crowd.

Ever since mathematicians identified Point A and Point B, they've been hanging around on the big grid of things that mathematicians call a graph. Sometimes mathematicians attach pairs of numbers to the points. Other times A and B find themselves appearing on circles, parabolas and all sorts of other curves.

One day A noticed something and asked B, "How come mathematicians always draw the same line through us? We have a variety of curves and shapes connecting us, but the same old line always passes through us."

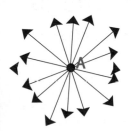

"You know you're right, A. Look, A, you can have infinitely many lines passing through you," B said, as it imagined all the lines passing through A.

"And so can you, B," A replied. "So why can't mathematicians use one of the infinitely many other lines? Are they too lazy?"

"No! no! no! you silly points," line AB (also called line L) exclaimed.

"We're not silly," A said. "We're curious and bored with you joining us all the time."

"Mathematicians have no choice," line L replied.

"What do you mean, no choice?" B asked. "Mathematicians are the ones who named us, came up with ordered pairs and this whole scheme of a graph. Mathematicians are the creators. They can do anything they like."

"That's where you've got it wrong," line L replied. "The mathematical system they created guides them, not the other way around. Yes, mathematicians *seem* to create and make up things, but those things exist. They have always existed, and mathematicians discover them as the system evolves before them."

"So why can't mathematicians use one of the infinitely many other lines? Are they too lazy?"

"Exists, systems, evolve. What is line L talking about?" A asked B.

"You've got me," was all B could say.

"I not only got you, but we're all stuck together for eternity," line L whispered.

"Why can't one of our infinitely many other lines also connect me and B?" A asked.

"It's an axiom," line L replied.

"So why not make a new axiom?" A asked.

"It's a rule," line L said.

"Well, just make up a new rule," A replied.

"An axiom is more than a rule," line L began to explain. "It's a truth for this system. Nothing else will work."

"Prove it," demanded B.

"That's just it. An axiom, also called a postulate, can't be proven," line L replied.

Point A was getting angry. "I've seen mathematicians continually come up with lots of statements which are then proven. This is just a statement out of thin air. To hold true, if I understand the process, it must be proven." Point A felt proud of the point it made.

"Well, that is true. If the *statement* is proven, it then becomes a *theorem*," line L tried to explain. "But there are a few which are true, very few I must say, that cannot be proven. These are accepted as truths."

"Well, with what you're saying to me mathematics seems to rely on faith," Point A countered.

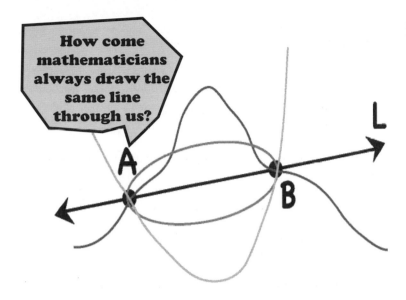

"In order to begin the mathematical system in which we now exist, you need to start with something. You can't start with nothing or a void, and expect to get something besides void. Right?" line L asked.

"I suppose," A and B hesitantly replied.

"See, mathematicians try to use as few axioms and undefined terms as possible," line L tried to continue.

"What do you mean undefined terms? Points A and B asked.

"You guys and I are undefined terms," line L replied.

"Now line L is telling us besides needing to accept that there is no other line besides L in our universe that contains both you and me, that we are also all undefined to boot," A said.

B appeared totally confused and was unable to utter a word for the longest time. Finally, B spoke up. "You know this whole mathematical system business seems like a religion. It requires a leap of faith to go along with it."

"No! no! no! one has to have logic—plain and simple," line L replied.

"That logic does not seem plain or simple," point B retorted.

Point A suddenly had a bright look about it. You know I've heard a rumor about new worlds."

"Has point A lost it?" line L said, looking at point B.

"Got me," was all B could say.

"Worlds where things are not as we are used to them," point A continued.

"Like what?" line L asked.

"Like curved lines. Like infinitely many such lines passing though two points," A said, as it gathered confidence and began to recall something it had heard.

"Have you listened to what I've been saying?" line L replied. "There are no other lines in Euclidean geometry that pass through you and point B—just me! Face it, you are stuck with me."

"Now it's all coming back to me," point A continued. " I heard a mathematician thinking out loud many years ago, and it wasn't Euclid."

"Who then?" line L asked.

"A fellow named George Bernhard Riemann. He made a different geometric mathematical system in which point B and I are not just stuck on you, line L. In fact, line L isn't in this system because this mathematical world exists on a sphere,

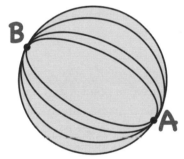

This illustration shows how infinitely many different lines (i.e. great circles) in spherical geometry can pass through A & B.

A great circle is a circle of a sphere that has the same center as the sphere.

and all its lines are actually great circles of the sphere. If you get the right sized sphere, so point B and I are opposite poles of the sphere, then there are infinitely many lines passing through both of us. Pretty creative, eh? He discovered it when trying to prove an axiom from Euclidean geometry."

Line L felt exasperated, and finally said, "Every time I get involved in a conversation with you two points, I end up dumbfounded. I recall last time you said a universe existed that could have triangles with angles totalling less than 180°."

"I'm sure," point A continued, "there are many mathematical systems. So what you were saying about mathematics having always existed, and mathematicians merely discovering its ideas rather than creating them seems so feasible."

"It does if you accept that there is more than one reality, and each has its own universe and truths," line L replied.

"Let me get this straight," B said to line L. "You're saying with each of these realities a mathematician really has no choice of what ideas evolve because they already exist, and the mathematician discovers them as they pursue ideas logically."

"That's how I see it," line L replied.

"Hmm...I'll have to give some thought to see if I accept all this," point B said, looking a bit confused at point A and line L. "I've had enough of this philosophical mathematical talk for now." Point B turned and chose another one of its infinitely many lines to hang out on, even though it knew line L would always be there passing through it and A.

The day the solids got truncated

Everything you've learned in school as 'obvious' becomes less and less obvious as you begin to study the universe. For example, there are no solids in the universe. There's not even a suggestion of a solid. There are no absolute continuums. There are no surfaces. There are no straight lines." **—Buckminster Fuller**

The slicer, as the plane was sometimes called, was in town. But why? What brought the slicer to Solidsville this time? The last time it was here it had transformed a cube into many pyramids by passing through it at different angles.

Slicer walked slowly into the general store. A small unsuspecting solid behind the counter asked, "May I help you?"

"Why, yes! thank you," the slicer answered in a pleasant voice, which made all the other solids in the store turn to look. They had only heard bad stories about the slicer. Could this be THE slicer they had heard about?

"I'm looking for the five Platonic solids[1]," the slicer continued. "I have a few surprises for them."

"They love surprises," the small solid replied. "Today they are all together at a meeting at the country club down the road."

"Thank you," the slicer replied and started down the road.

As slicer walked, the solids scattered and cut a wide swath for it to pass.

*   *   *

Meanwhile, at the country club the Platonic solids lay basking by the pool. It was one of those lazy summer days, and they had decided to take a pool break before getting back to the business of finding the best and most efficient way to pack solids in a rectangular container.

"Guess what, regular hexahedron," asked tetrahedron addressing cube. Tetrahedron was very formal and could never bring itself to call the regular hexahedron by its nickname, cube. "I have been thinking about mixing certain solids when packing them."

"I thought we had all agreed to take a break from work and just relax. Let's not talk shop," cube replied, half asleep.

At that moment slicer entered into the pool area. The Platonic solids were suddenly all attentive, except cube, who had dosed off.

"Whaaaaaaat are you doing here slicer?" tetrahedron asked in a timid tone.

"No fooling around," cube replied, its eyes still closed. "You can't frighten me with that old joke pretending you're the slicer."

"It is not a joke," slicer replied.

At which cube leaped off of its lounge, recognizing slicer's voice.

"What do you want?" cube asked, trying not to appear scared.

"I've come to truncate you," slicer replied.

"We are really fine just as we are…. I don't plan to do any traveling. …I don't need a trunk…. As you know each of my faces are identical squares, all my angles are right, my edges are all congruent and I have eight vertices — I certainly don't want an elephant trunk for a nose. I am just fine," cube stammered.

As slicer walked, the solids scattered and cut a wide swath for it to pass.

"Exactly my point," slicer began. "You Platonic solids could use a change. It's about time you experienced a new look. Just relax," slicer continued.

"No, please, not me," cube pleaded.

"Don't be frightened, cube. You know it doesn't hurt, and it isn't permanent. It is like getting a new hair style," Slicer reassured cube. And with that comment, Slicer, the plane, moved so swiftly that all 8 corners of cube were cut off. The other Platonic solids were in a state of shock. They couldn't believe how cube looked. They couldn't help it, but they began to laugh. Slicer turned toward icosahedron, and they all became quiet. "Don't you think cube looks intriguing now?" slicer asked. "Oh yes," they all replied.

"Well then, let's give all of you a new look too." And with that comment, slicer transformed the other four Platonic solids into their truncated versions, so each corner appeared to be cut off and flattened.

"Now that's what I call a geometric facelift," slicer laughed as it walked away leaving the 5 Platonic solids to piece themselves together again.

---

[1]The ancient Greek philosopher/mathematician, Plato proved that there are only five convex 3-dimensional solids whose face are congruent polygons — the tetrahedron whose four faces are equilateral triangles, the regular hexahedron (cube) whose six faces are squares, the octahedron whose eight faces are equilateral triangles, the dodecahedron whose twelve faces are pentagons, and the icosahedron whose twenty faces are equilateral triangle.

The diagrams below illustrate these five Platonic solids and their truncated forms.

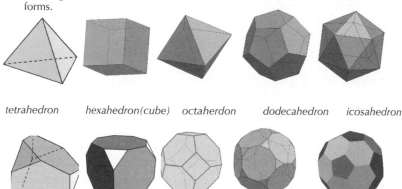

tetrahedron   hexahedron(cube)   octaherdon   dodecahedron   icosahedron

What's the point of a point?

...Nobody knew then that there could be space. ... in reality there wasn't even space to pack us into. Every point of each of us coincided with every point of each of the others in a single point. —Italo Calvino

One day *point, line, plane* and other geometric objects were gathered in a part of *space*, just hanging out..

"Did you see that beautiful octagon that some person drew on me the other day?" *plane* asked the group.

"Yes! Very cool," *triangle* replied, while lounging on another plane.

"Yes, that was impressive," *line* added. "But how did you like me decked out as the number line yesterday?"

"Very stylish," *space* interjected.

"I was used in a neat problem involving me, a circle and a square," *triangle* said, compelled to share what it had been doing.

Suddenly point burst into tears and began to cry and cry and cry. The other objects immediately turned their attention to *point*.

"Why are you crying?" asked *line*.

"I feel totally useless," *point* said sadly. " People cannot see me. I have no dimension. They only know where I am when a big fat dot is placed on me. It's humiliating. What's the point of me?" *point* asked, tears streaming down its face.

"Don't cry," *space* pleaded. "You are useful. Even though you are so very tiny and people can't see you, they know you're there."

"There are many minute things humans cannot see, but they know they are there," the *line* added.

Suddenly point began to cry and cry and cry.

"What?" asked *point* sniffling.

"A single cell, a molecule, an atom, an electron, a neutron, a quark, for example. You're like one of those—invisible to the human eye," *triangle* said. "When clustered you make things, just like molecules come together to form objects."

"What do I form?" *point* asked.

"You form us. You make lines, planes, triangles, squares, circles …even space is defined in terms of you—the set of all points," *plane* answered.

"Yes, but a molecule, a cell, an atom all exist. They can be seen with a microscope or by experiments. What about me?" *point* asked, crying again.

"You exist in the imagination. We all do. That doesn't make us less important or less valuable," all the objects now declared in unison.

"If a single point were missing from my line, I wouldn't be a line. And just think, one number would have lost its location on the number line," *line* added.

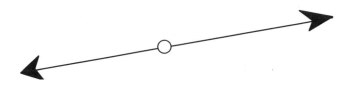

"If one single point, say a vertex of my triangle were gone, I would no longer be a triangle," *triangle* demonstrated.

"If one point left *space*, *space* would no longer be defined," *circle* said and added, "If my center were gone, I would be no more. We need you *point*. You are our essence. Without you we don't exist."

Point stopped crying and a happy glow came over its face. *Point* finally understood the point of *point*.

Imagine i

The imaginary number is a fine and wonderful recourse of the divine spirit, almost an amphibian between being and not being.

—Gottfried Wilhelm Leibniz

i sat alone feeling sorry for itself. It felt it just didn't fit in. i had never been invited to join a set or used to solve a problem. In fact, i was totally ignored and felt completely useless. i was envious of π, who had been an active member of the world of numbers since the Babylonians tried to pinpoint its value over three thousand years ago. Ever since then π has teased mathematicians with its elusive decimal value and invaluable work from early circle problems to probability to modern computing. Yes, π was totally contented, especially when it was finally given its official symbol name[1] in the 17th century.

"When is my day coming?" i wondered.

A set of negative numbers came over to where i was sitting. "So you're just moping around as usual. Not yet involved in any math problems, eh?" –3 taunted. "I'd say i is a totally unnecessary number," –3 taunted.

"That's for sure," the other negative numbers said and nodded. "i should not even be allowed in number world. It's just a letter with no number work."

"Not so fast," 1 interrupted. "i has been used as a subscript."

"But that's not a number. In fact, that's not much different than the variable x," −3 added.

"Don't worry i. Be patient," 1 encouraged i.

"I can remember," 1 continued, glaring at the negatives, "when there were no negatives, and they were all nothing, oops I mean nonexistent, in the real world of people. Negatives are actually less than nothing, or should I say less than zero. There was a time when mathematicians didn't know about negatives, let alone know how to use them. I recall Blaise Pascal, that French mathematician, saying *'I know those who could not understand that to take four from zero remains zero.'* At that time, he didn't even realize negatives existed."

"I know, I know," replied −3. "I'd like to forget those days."

"So why tease i?" 1 asked. "I'm sure its time will come."

"But the symbol i for a number?" 0.5 asked. "What could it mean?"

"It's no stranger than the symbols π and e," 1 explained.

"To me it seems more like a variable than a number," −1 chimed in.

At that precise moment, the famous variable x rushed toward −1, shouting. "It's happened! "It's happened! Mathematicians[2] have used me and you in an equation.

"So," said −1. "It's not the first time. Recall when the equation x+1=0 was used to solve some problem? Mathematicians were sure surprised when they didn't have one of their known numbers to solve that equation," −1 chuckled. "And so the negative numbers were finally made official and useful numbers."

"It's different this time. You and I were put in the quadratic equation, $x^2+1=0$. When they solved it, $x^2 = -1$, then they found x to equal $\pm\sqrt{-1}$. They had no idea what this meant. There were no numbers to represent $+\sqrt{-1}$ and $-\sqrt{-1}$. "

"Well," 1 asked anxiously, "what happened."

"They said the solution is imaginary", x replied. And decided to call $\sqrt{-1}$ by the name i.

Suddenly i perked up. "You mean they chose me to stand for $\sqrt{-1}$, and said I am to be used for this imaginary number?"

"Yes!" shouted x.

"See," 1 said. "Didn't I tell you some work would come up for you."

i felt elated and sad at the same time.

"What's the matter?" 1 asked, sensing i's melancholy.

"I'm imaginary," i replied. "I wanted to be real."

"You're more and better than real," x quickly replied.

All the numbers turned toward x questioningly, and asked, "Better than real numbers?"

"Let me explain," x said, realizing it had offended the others. "i is called imaginary, but it is a number that helps encompass all of you numbers—counting, whole numbers, integers, rational, irrational, transcendental, algebraic. You name it, you are all in the same set now."

"Impossible," 1 said. "How can that be?"

"It's all on account of i," x explained. "You are all part of the complex number set."

"Complex!" all the numbers shouted, including i.

"Mathematicians now have declared that any number written in the form a+bi, where a and b are real numbers, and **a** represents its real part and **bi** its imaginary part, is a COMPLEX NUMBER."

"So how am I to be written?" asked 1.

"1+0i, " x replied.

"What about me?" $\sqrt{5}$ asked. "And me, and me, …"   All the numbers wanted to know their complex form.

The barrage of questions exhausted x.

i cleared its throat, and finally asked, "So my complex form must be 0+i."

"Correct," x replied, breathing very rapidly.

i was so enthusiastic it began to dance and sing, "I'm a number. I'm imaginary. I'm complex, I'm useful!…"

x finally caught its second wind, and said "There is so much more math stuff now because of you, i."

"What do you mean?" −3 asked.

i sat alone feeling sorry for itself. …, i was totally ignored and felt completely useless.

"For example," x began, "because of i, all solutions of a polynomial equation can now be found. In the past, for example, the equation $x^3 = -1$ had only -1 identified as the answer to this problem, but Gauss' work[3] proved it has 3 solutions."

"Really?" −1 asked, feeling a bit dejected.

"Really!" x replied. "The solutions are -1, $(1+i\sqrt{3})/2$ and $(1-i\sqrt{3})/2$. Cube each and you'll see how their cubes make -1."

"There is much more to share…," x started to continue, but was interrupted by 1.

"Stop, we've heard enough. i you are definitely one of us now, or should I say we're one of each other as complex numbers. "

"But I forgot to mention a really important part," x shouted over all the number buzz. "$e^{\pi i} + 1 = 0$ … it's called Euler's identity.[4]"

All the numbers turned and looked, and were astonished to find five of the most famous numbers all in one neat mathematical sentence…all because of i. The numbers were awestruck. Then they clapped thunderously as they turned toward i.

"Now," i said, "I really feel **alIve**. What will those mathematicians dream up next?" i walked away with a spring to its step.

---

[1] The symbol π appeared for the first time in 1652 in the work of William Oughtred, and in *Synopsis palmarioriJum matheseos* by William Jones .

[2] In 1572, Rafael Bombelli is recorded to have done the first calculations using imaginary numbers.

[3] The fundamental theorem of algebra, stating that *"every polynomial equation of degree n with complex number coefficients has n solutions over the complex numbers,"* was proven by Carl Gauss in 1799.

[4] Euler introduced this identity in his book *Introductio* in 1748.

The attack
of the locus

When you have eliminated the impossible, whatever remains, however improbable must be the truth.

—Sir Arthur Conan Doyle
The Sign of Four

Point was relaxing in its zero dimension, taking in some sunshine. All of a sudden hundreds— no, thousands! no, millions! no, trillions! — in fact, infinitely many points from nowhere converged on its place. Then they simultaneously jetted out from where Point sat, and stopped suddenly precisely 5 inches away. Point felt a sign of relief … no longer pinned by the weight of a mountain of dimensionless points.
 As Point looked around, it saw an amazing sight. A solid mass of points now surrounded Point, appearing as a line — well, not exactly a line — for this line had no beginning and no end and was curved.

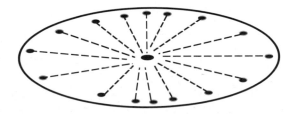

"What are you?" Point asked.

"Can't you see, we form a circle?" they replied.

"You're so beautiful! Your curve is so constant, always 5 units from me," Point replied in awe of them.

"Thank you, but we can be 6 units away, 1/2 a unit away, even a 1000 units away. In fact, any distance away, and we will still be a circle." As they spoke, they moved in unison 6 inches away, then moved in to 1/2 inch, and then faraway so their voices dimmed as they approached 1000 inches, and then back to 5 inches. "We are a locus of points equidistant from you."

"Locusts? Are you insects?" Point asked.

"No! not locusts. L-o-c-u-s," they all spelled.

"So a locus is a circle," Point said with a smile.

"No! not locusts.
L-o-c-u-s"

"Not necessarily," they replied. "There are many other things formed by us. It all depends on what conditions are placed on us. For example, we are a circle if we are told we must be equally distant from you or some other point and lie on one plane. But if we are told just to be equally distant from you, we won't form a circle."

"Well, what do you form?" Point asked puzzled.

"You think about it. What would all the points around you form if the only condition was that we must be 5 inches from you? Remember that includes about and below you."

"I see! It's a sphere. Give me some other locuses to work on," Point asked.

"Please, the plural of locus is not locuses, but loci," they answered, a bit miffed.

"Oh, I'm sorry, I just didn't know," Point replied.

"Not a problem. We just like everyone to get things straight. So let us ask you to discover some other loci. Just think about the conditions carefully."

1. Find the locus of points on a plane that are equally distant from these two points.

2. How about finding what the locus is, if all the points are equally distant from the two fixed point but not confined to a plane, but can be all over space?

3. What about the locus of points on a plane whose distance from these two fixed points always totals 10 inches? These two fixed points happen to be 4 inches apart.

4. What is the locus of all points on a plane that are equidistant from a given point and a given line?

5. What is the locus of the points traced by a point that is on the boundary of a circle that is rolling along a straight line?

"These problems seem difficult. I've never even done a locus problem before," Point complained.

Sure you have. You just did one before with your answer sphere," the locus points replied. "Don't be a chicken. You're the center of our attention for awhile. At least think about them, and take a stab at them."

"Sure why not." Point now said sounding confident as it sat in the center of the circle thinking about the questions.

"We'll leave you the answers below." And with that the locus points dashed off like a swam of locusts.

---

answers

1) The line that is the perpendicular bisector of the two fixed points.

2) The plane that contains the perpendicular bisector of the two fixed points, and every point on this plane is equidistant from the two fixed points.

3) An ellipse whose foci are the two fixed points.

4) A parabola whose given point is called its focus and the given line is called its directrix.

5) a cycloid.

Permutation & Combination have it out

It has been pointed out already that no knowledge of probabilities, less in degree than certainty, helps us to know what conclusions are true, and that there is no direct relation between the truth of a proposition and its probability. Probability begins and ends with probability.

—John Maynard Keynes

"Stop,"shouted Permutation. "Don't put that can back there on the shelf!"

"Why not?" asked Combination.

"I had the five cans arranged in a particular order," Permutation replied.

"That's silly," Combination said. "Why does it matter how these cans are arranged? These five cans are simply five cans. What kind of control freak are you? Who cares if you arranged them by color, alphabetically, size, et cetera? "

"I care," Permutation replied with an angry and exasperated voice. "It makes a difference to ME! Each arrangement is a permutation, which is what I am all about."

"You are just so fussy," Computation rebutted again.

"We are just different. To me the order of the arrangement is always important," Permutation said empathically.

"I beg to differ," Combination said in a polite tone trying not to lose its temper.

"Look," Permutation continued. "If six people line up to get tickets for a movie, order is important to each of them, right?"

"Now, you are getting tricky," Combination said. "If I had five cans of the same tuna on a shelf, is order important? Does it make any difference which can is on the top or on the bottom?" Combination asked with a smile on its face.

"Now look who's getting tricky," Permutation replied.

"You know it all depends on the problem," Combination added. "And furthermore, I am a very important concept in probability.

"What kind of control freak are you? Who cares if you arranged them by color, …, etcetera?"

"Yes, I know, I know. But, I'm just as important, " Permutation shot back.

"Let's not argue," Combination said trying to be conciliatory. "Let's agree that we are just different."

"That's what I said in the beginning," Permutation declared.

"And by the way, stop calling me a control freak," Permutation added.

"Let's consider another situation. How would you approach this problem?" Combination asked. " Suppose you toss three pennies in the air, how many ways can they land?"

"Why that's a permutation problem," Permutation said.

"Darn. Let me think of something else that is a combination problem," Combination replied.

"Why don't you see how many different ways three people can shake hands? The order that two people shake hands is not important because they are both shaking hands simultaneously. Therefore this is a combination problem," Permutation said helping out Combination.

"You're right. Thanks. That's nice of you to think of that one," Combination said. " So if a doctor, a teacher and a lawyer were blindfolded and taken into a room, what is the probability that the doctor and lawyer will be the first to shake hands?"

"What if they don't want to shake hands because the lawyer thinks he/she will be entering into an agreement?" Permutation said.

"Don't be silly, Permutation. Just suppose that all three want to shake hands. What's the probability that the doctor and lawyer will shake hands first?"

"Well, since there are only three combinations of how they can shake hands, the probability is 1/3," Permutation answered.

"See, knowing combinations is an important and useful concept," Combination said proudly.

"Let's go back to your problem with the pennies," Permutation proposed. " What is the probability that the 3 pennies all land as heads."

"I feel a little foggy today," Combination replied. "Why don't you explain how to find the answer?"

"Okay. In this case we must find all the possible ways three pennies can land. That will tell us how many outcomes are possible, and in this case the order in which they fall is important," Permutation explain.

"Just a minute, what's the difference when you get two heads and a tail?" Combination countered.

"It's a big difference because you might first have a head and then a head and finally a tail land. Let's write that as HHT. But they may come down in the order THH or HTH. So there are 3 possible outcomes for 2 heads and a tail. And that's not all. What if 2 tails and a head come up. How many of these outcomes are possible?"

"Following what we just did it would be 3 outcomes with TTH, THT, HTT," Combination immediately said. "I guess my head is clearing up?" Combination joked.

"So far we have 6 possible outcomes. Have we got them all?" asked Permutation.

"You know we haven't," Combination replied.

"Well, what have we missed?" Permutation asked, testing Combination.

"What about if they all land as heads. That is HHH,. Or all as tails. That's TTT. So there are 8 possible outcomes." Combination explained.

"Excellent," Permutation complimented Combination. "So the probability of getting 3 heads, HHH is..."

"Is 1/8," Combination answered not giving Permutation a chance to give its answer.

"Well, let me add that finding the permutations for tossing coins in the air are easily given by using the Pascal triangle*," Permutation added.

"There you go again,

---

*ANNOTATION_____

There are 4 combinations for 3 coins tossed. These are: three heads; two heads & a tail;   a head and two; and three tails.

From the story we see the permutations are 8.

  1 for HHH
  3 for HHT HTH THH
  3 for HTT THT TTH
+ 1 for TTT.
  8

```
        |           —row 0
       |  |         —row 1
      | 2 |         —row 2
     | 3  3 |       —row 3
    | 4  6  4 |     —row 4
     .  .  .  .  .  .
```

Notice these numbers correspond to the 3rd row of the Pascal triangle.

name dropping," Combination countered. "Did you know that the Pascal triangle is related to Newton's binomial formula?"**

$$1 \qquad = (a+b)^0$$

**ANNOTATION _____

$$1\ 1 \qquad = (a+b)^1$$

$$1\ 2\ 1 \qquad = (a+b)^2$$

$$1\ 3\ 3\ 1 \qquad = (a+b)^3$$

$$1\ 4\ 6\ 3\ 1 \qquad = (a+b)^4$$

$$\cdots\cdots = (a+b)^n = \binom{n}{o}a^n + \binom{n}{1}a^{n-1}b^1 + \binom{n}{2}a^{n-2}b^2 + \ldots + \binom{n}{n-1}a^1b^{n-1} + \binom{n}{n}b^n$$

Newton's binomial formula

"Now who's name dropping," Permutation replied.

"We'd been getting along so well, until we started our one upmanship," Combination added. "Let's call a truce."

"Yes, let's," Permutation agreed. "Especially since we are both very important in various probability problems."

"And we both have our own formulas for calculating us," Combination said, having the last word as it walked into its probability section of a math book.

Composites take on the primes

Six is a number perfect in itself, and not because God created the world in six days; rather the contrary is true. God created the world in six days because this number is perfect, and it would remain perfect, even if the work of the six days did not exist. —St. Augustine

"**W**hat do you think you are doing here?" 5 questioned 6. "You are not prime. Only primes are allowed at this set gathering."

"Where should I go then?" 6 asked in a pleading voice.

"I don't know. I just know you don't belong here. Go hang around with 8 and 9. Like you, they are not primes," 3 sneered at 6.

6 had become increasingly annoyed at the rudeness of the prime numbers. They often bragged about how special and essential they were because products of them formed the rest of the counting numbers, and because they were the major players in the Fundamental Theorem of Algebra[1]. 6 decided to seek out 1 to talk about the prime numbers, since 1, being the first number, was respected by all numbers.

"Yes, I understand how you feel, 6. They do get a bit snooty," 1 said consolingly. "We are all part of the whole numbers. We should try to get along. They just like to hang around in their own subset. Why don't you go and form your own subset. There are enough of you. What with you and 4, 8, 9, 10, 12, 14, 15, 16, and on and on."

"You mean all the rest of the whole numbers? The ones that are not prime?" 6 asked.

"Well, all except me and 0," 1 replied. "I am not one of you because I have no prime numbers as factors. My only factors are 1s. And 0 has every number as a factor, since 0 times any number is 0. So we are in a class of our own."

"I see," 6 replied. "So my subset would have only numbers which have at least two prime factors. They could be the same prime number. For 9 it would be 3x3. Right? But the important thing is we're all composed of primes. We are composites of primes. That's it!

"Go for it," encouraged 1.

"We will be called the composites," 6 shouted excitedly. "My subset has a name, and there are as many of us as there are primes, probably a lot more. ... Oh, I feel so much better now," 6 told 1, and off it walked with a spring to its step.

* * *

Seeing 5 and 3, 6 hollered, "Hey, 5 and 3. Bet I can beat you at a game of rectangles?"

"Fat chance you could beat me at anything," 3 replied. "How do you play?"

"You use your different factors to make different shaped

rectangles that would have, in your case, area 3. Whichever of us makes the most is the winner," 6 explained.

3 immediately formed a 1 by 3 and a 3 by 1 rectangles, since these each had area 3.

Then 6 went to work and made a 1 by 6, a 6 by 1, a 2 by 3 and a 3 by 2. "

"I win!" 6 shouted.

"Okay, okay, don't rub it in," 3 replied.

"In fact, I am certain I can beat any prime number at a game of rectangles," 6 boasted. "Actually, I am certain you can't beat any of us composite numbers," 6 said proudly.

"How is that possible?" 5 asked.

"That is because we are composite numbers," 6 replied.

"Composites? What are they?" 5 asked.

"We also happen to be a subset of the whole numbers. Each composite has two or more prime numbers as factors, which explains why any of us can outmaneuver any prime, no matter how big, at a game of rectangle?"

"How is that possible?" 5 asked again.

"Well, how are prime numbers defined?" 6 asked 5.

"You and every other number know our definition. Each prime number is a counting number whose only factors can be 1 and itself," 5 replied.

"Since each of you has only two factors, how many different rectangles can be formed?" 6 continued questioning 5.

"And by the way, besides being a composite number I am also a perfect number"

"Why just two, and actually these two happen to be the same rectangle, but just flipped around 90 degrees," 5 answered.

"Exactly," 6 responded. "Since all composites have to have at least two prime numbers as factors, they will form at least four rectangles, counting the the ones that are flipped around 90°. So naturally we will always win at the game of rectangles."

5 felt disheartened, and began to walk away, when 6 said, "Just wait. Don't walk out on me like that. I am not done speaking with you."

5 stopped and turned toward 6 saying, "Now what do you have to say?"

"That you are not always prime," 6 dared to say.

"Just one minute," 5 said. "You composites may win all the game of rectangles, but I do know we primes are always prime."

"Not if we make one slight change to your definition. After all definitions can change depending on what area or part of math one delves in," 6 replied.

"Prove that by giving me an example," 5 challenged 6 .

"No problem," 6 replied. "Consider Euclidean geometry versus spherical geometry. In Euclidean geometry a line is described as a 2-dimensional set composed of points that are aligned straight and stretch out infinitely in opposite directions. But in spherical geometry, a line is defined as a great circle[2] of its sphere. So if we change the definition of prime numbers to not be counting numbers, but rational or even complex numbers, no primes exist!" 6 declared.

"What's wrong with you? That is the most ridiculous statement I've ever heard. No primes. Hah!" 5 answered 6.

"Let me illustrate what I mean. If the factors of a number can only be whole numbers, then sure you are a prime number. But, if your factors can be fractions, i.e. rational numbers, then you have lots of factors," 6 began explaining.

"Like what?" 5 demanded.

"Like a 1/2 and 5/2' since 1/2 x 5/2 = 5," 6 pointed out. "In fact there are infinitely many possible rational factors for you."

"And when you use complex numbers, what are my factors?" 5 asked.

"They are (2+i) and (2-i). Multiply them together and you'll see their product is 5. And by the way, besides being a composite number I am also a perfect number[3]." And with that comment 6 said "See you around," and sashayed away.

---

[1]The Fundamental Theorem of Arithmetic, also know as the Unique Prime Factorization Theorem, states that any counting number greater than 1 can be written as a unique product of prime numbers. The order or arrangement of the prime factors is not considered in determining how many ways the number can be factored. Here is an some example: $12 = 1 \cdot 2 \cdot 3 \cdot 2$ is considered the same unique factorization as $1 \cdot 2 \cdot 2 \cdot 3$ or $1 \cdot 2^2 \cdot 3$, etc..

[2]A great circle of a sphere must be a circle of the sphere with the same center as the sphere.

[3]A number is a perfect number if the sum of its proper divisors total the number. A proper divisor is a factor of the number other than itself. For example, since $6 = 1 \cdot 2 \cdot 3$ and $1 + 2 + 3 = 6$, **6** is a perfect number. 28 is the second perfect number. The sixth perfect number, 8,589,869,056, was discovered in the 16th century by Italian mathematician Pietro Cataldi.

How the points
of a plane
got their names

With me everything turns into
mathematics.
—René Descartes

THE TIME — the beginning of the 17th century.

THE PLACE — France.

"Did you hear?" shouted a point from the plane of a sheet of paper.

"Hear what?" another point from the paper asked.

"That we are no longer anonymous points of a plane. We now have names!" the first point shouted gleefully.

"How can that be?" the second point was skeptical. "Then what is my name?" it demanded.

"Today you are 1,3," the first point answered.

"What kind of a name is 1,3? You mean to tell me that my name is made up of numbers? Well then, what is your name?" 1,3 said, firing a set of questions at the first point.

"I'm 7,2," the first point replied.

"Your name is also numbers?" 1,3 said, surprised.

"Yes, every point on the plane has a unique pair of numbers

as its name," 7,2 answered.

"Who wants numbers as their name? I sure don't!" 1,3 replied angrily.

7,2 said, "I don't mind. It is better than no name. I'm tired of being called point, dot, place or location with no name to distinguish you from me. Now at least we each have our own identity thanks to René Descartes."

"What kind of a name is 1,3? ... my name is made up of numbers? "

"Descartes? Who's this Descartes? What right does he have to name each of us. Are we related?"

"No! He is a well known mathematician who came up with this method of naming each and every point of a plane. Recall how many of us there are on a plane — infinitely many — all crammed together. Yet he has come up with a way to give each of us a unique name — to tell people exactly where each of us is located without having to point a finger, especially since it is impolite to point. We don't ever have to be called by the same name again. I am satisfied even if my name is made up of numbers," 7,2 declared as it raised its voice. "Which of us has not been called point or dot? We have never had our own names before. We will never be confused with another point. I finally have an identity!" 7,2 shouted at the top of its voice.

1,3 finally began to realize the importance of what 7,2, was saying. "I see! I guess it is good to have one's own name. It is just hard to adjust to having a name made of numbers. What's the difference between my name 1,3 and 3,1?"

"The order of the numbers makes the difference," 7,2 replied. "The first number refers to the x-axis number and the second goes along with the y-axis number. 1,3 and 3,1

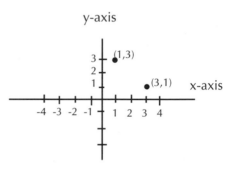

are at different locations on the plane, so their names are different."

Just then a voice from the intersection of the two perpendicular lines shouted, "I feel real special!"

"Why are you butting into our conversation? What makes you think you are so special?" 1,3 asked.

"My number name is 0,0, but I am also called the origin," 0,0 replied with a big grin.

"Don't feel too smug," 7,2 replied. "You're just lucky the two perpendicular lines, called x and y axes, were drawn passing through you, but someone else can draw their own set of lines on the plane, and they could pass through me. Then I would be 0,0 and you would have another name. It all depends on where the axes are drawn. That's how our names are organized. And since a plane is infinite in two directions, it has no predetermined middle. It can span out from wherever the two axes are drawn."

"Don't confuse me 7,2 just when I was getting accustomed to my new name," 3,2 said. "Let's get used to this set of axes before playing musical names. I'm fine with that point being the origin for a while. It doesn't bother me, especially since I now have my own name."

"Well, that's a switch. I am glad to hear it. Here's to individuality! Here's to Descartes!" 7,2 shouted. The points on the plane all yelled their new number names together, and a babble arose from the plane.

The 2nd Millennium Awards

I have often admired the mystical way of Pythagoras and the secret magic of numbers. —Sir Thomas Browne

The crowds outside were growing and growing. The theater was already jammed. After all, tonight was not just a decade celebration, a bicentennial, or even a centennial. Tonight was the night a millennium of mathematical ideas would be recognized at the MMA ceremony — Mathematical Millennium Awards. Numbers had come from the far reaches of the complex number plane. Not one number wanted to miss this once in a 10^3 happening.

"Look who just arrived!" shouted the outdoor interviewer. "The Pythagorean Theorem equation, one of tonight's presenters who has been famous since the 2nd millennium B.C.. "What do you think of this millennium's nominees?" the interviewer asked.

"I'm amazed at how mathematics has taken off over these past 1000 years. The awards for this millennium are going to be very difficult for the judges," the Pythagorean theorem replied.

"You'll be presenting *the best equation* award, won't you?" the interviewer quickly asked. $a^2+b^2=c^2$ didn't waste a nanosecond taking the lime light. A big crowd of numbers turned towards the Pythagorean equation, and it reveled in the attention. "Thank you, Thank you. It's amazing how over the centuries so many uses for me have been discovered," the equation began to brag. "First in architecture with the Pyramids and Parthenon, and then the solving of the quadratic…" "Yes, Yes," the interviewer interrupted. "We'll be looking forward to your presentation on stage. For those of you who have just tuned into the show, let me mention once again that the awards are kept in a hypervault until the awards ceremony. After all nominees for all categories have been announced, members of the guild of mathematics will cast their votes."

"Passing before us now are a few nominees for *best equation for this millennium*. Why, there is Einstein's $E=MC^2$, nominated for its role in relativity theory. Chatting along side is Newton's gravitational formula: $f=(Gm_1m_2)/r^2$. And just behind is the famous expression for e, we see $e=\lim_{n\to\infty}(1+1/n)^n$. But look! There is the equation that holds five of the most famous numbers: $e^{\pi i}+1=0$. What a treat to see all these equations gathered together."

"Look! Look! there are the 4 conic sections—parabola, hyperbola, ellipse and circle — who will present the awards for *best curves for an equation*. Hyperbola is stunning this evening decked out in a pair of silver axes approached by its asymptotic curves in gold. Circle is escorted by one of its diameters. Parabola is here with its focus and axis of symmetry. And ellipse has come with its two foci straddled on either side by the major and minor axes. These superstars of mathematics said in unison, "We look forward to comparing curves with the nominees— cycloid, Daubechies wavelets, sinusoidal curves, limaçon, and the equiangular spiral, who could receive a second consecutive millennium nomination— but you'll have to excuse us, we have pre-award appointments with our graphers."

"Hyperbola is stunning this evening decked out in a pair of silver axes approached by its asymptotic curves in gold."

The interviewer was a bit disappointed, but the three impossible problems of antiquity, *squaring the circle, trisecting an angle* and *duplicating a cube* (all with just a compass and straightedge) caught his attention. "Tell me," the interviewer asked *trisecting angle,* "how do you three remain so popular after 200 years?"

"We have never lost our fascination," *duplicating cube* answered. "It's more than that," *trisecting angle* added. "There are all sorts of stories about how we came about, what with the Oracle at Delphi and the gods," *duplicating cube* said. "I believe it is because people are in awe of the impossible," *squaring the circle* interjected. "They cannot resist thinking something was overlooked, even though they know we have been proven impossible. I am certain Fermat's Last Theorem will remain popular for ages. There will always be some mathematicians looking for an easier method of proof."

Thank you for talking with us," the interviewer said, as they proceeded toward the entrance. "Those old timers are going to give out the awards for *solving the best of the "unsolved" problems.* The competition is stiff this millennium what with *Fermat's Last Theorem* solved by Andrew Wiles, Euler solution of the *Königsberg bridge problem, the four color map problem* by K. Appel and W. Haken, let alone *the Parallel postulate problem* with the work of Lobachevsky, Bolyai, Riemann. Then there is the controversy about whether to allow *Kepler's sphere-packing problem* solution by Thomas Hales to be a nominee. Its solution has not yet passed all the referees. As we know, each millennium awards has its problems and the 2nd millennium AD is no exception."

"Look! Here comes Euclidean geometry stretched out on a plane with the book *Elements* along side the other undefined terms, point and line. "Who are the contenders for *the best new geometry?"* the interviewer yelled at Euclidean geometry. "Those strange geometries, who tried to displace my 5th Postulate. That fractal geometry which insists it describes things better than I can. Oh yes, Descartes and his

Cartesian Coordinate system in analytic geometry. Imagine proving things using algebra rather than brute logic. It seems so mechanical. And finally topology, that rubbery geometry, that stretches and pulls and distorts things. Can you imagine! It is hard for me to fathom sharing a domain that I had to myself for so many hundreds of years. I must get inside and relax. All this talk has made me nervous." So saying, Euclidean plane geometry slipped away.

"Oh great!" the interviewer shouted as he caught a glimpse of last millennium's best quirky number, the $\sqrt{2}$, who was decked out with an infinitely long train of endless non-repeating decimals. "I understand the nominees for your award are $\sqrt{-1}$ also known as i, W.R. Hamilton's quaternions , fuzzy numbers, Cantor's transfinite numbers,and the transcendental numbers," the interviewer said. "Well, you may understand them," $\sqrt{2}$ added, " but they thoroughly confuse me. I have no idea how mathematicians came up with such crazy numbers. Imagine! an imaginary number, 4-dimensional numbers, numbers that don't define fixed quantities, numbers for designating infinite amounts, and the transcendental numbers — forget them."

"But you were not exactly considered conventional by the Pythagoreans," the interview challenged $\sqrt{2}$. "I don't want to talk about my past," $\sqrt{2}$ said, and hurried away.

"I must say, the $\sqrt{2}$ actions seem very irrational," the interviewer added before pointing out three of the nominees for *curves with the best special effects*. "Look over there everyone. We have the equiangular spiral nominated for its roles in the golden rectangle, the golden triangle, and in natural phenomena. Just behind it is the double helix for its incredible appearance in DNA, and rolling right along side is the cycloid nominated for its tracking of the movement of a point on a wheel."

"I must say this has been an incredible millennium," the interviewer reflected. We've only dealt with a few categories, yet the concepts and ideas seem overwhelming. Just look at the line of celebrities waiting to enter! —For *best number in a problem's solutions* I see the transcendentals' number *e* for its phenomenal performance in half-lives and its part in keeping interest rates under control; *quaternion,* the 4-dimensional number, for its work in computer graphics; *i,* the imaginary number, for its undaunting work in complex number problems. Lastly there is *phi,* the golden mean, for its role in the Fibonacci sequence. The nominees go on and on."

"And what about the category, *most clever idea leading to innovative computing techniques?* There are Napier's logarithms and their part in Napier's bones; and look there is the *Universal Turing machine* and its idea of computability, and here come Babbage's analog and difference engines. This is going to be an amazing show! ... I can't believe it! There are Cantor's *transfinite numbers,* nominated for *best mathematical work not recognized during the mathematician's lifetime.* And also for that award I see the *Saccheri rectangle* for Saccheri's work on the Parallel postulate, even though Saccheri himself may not have recognized the significance of his work."

"This millennium has many new categories including new technologies. … For example, look what's coming. Some of the nominees for *most innovative computer*— the quantum computer with its particles, the molecular computer with its DNA and the optical computer clutching to its laser beam."

"Yes, this is indeed going to be a phenomenal show. We have only seen a fraction of the nominees. Who will you vote for? Will we run past the program's allotted time?" With these questions the interviewer finally broke for a commercial.

Author checking out the concrete "tetrapods" in Santa Cruz, CA.
Photograph by Joan Newman.

Theoni Pappas is passionate about mathematics. A native Californian, Pappas received her B.A. from the University of California at Berkeley in 1966 and her M.A. from Stanford University in 1967. She taught high school and college mathematics for nearly two decades, then turned to writing a remarkable series of innovative books which reflect her commitment to demystifying mathematics and making the subject more approachable. Through her pithy, non-threatening and easily comprehensible style, she breaks down mathematical prejudices and barriers to help one realize that mathematics is a dynamic world of fascinating ideas that can be easily accessible to the layperson.

Her math calendars and over 15 math books appeal to both young and adult audiences and intrigue the "I hate math people" as well as math enthusiasts. Three of her books have been Book-of-the Month Club™ selections, and her Joy of Mathematics was selected as a Pick of the Paperbacks. Her books have been translated into Japanese, Finnish, French, Slovakian, Czech, Korean, Turkish, Russian, Thai, simplified and traditional Chinese, Portuguese, Italian, Vietnamese, and Spanish.

In 2000 Pappas received the Excellence in Achievement Award from the University of California Alumni Association for her work on books and other innovative products that deal with mathematics.

In addition to mathematics Pappas enjoys the outdoors, especially the seashore where she has a home. There she bicycles, hikes and swims. Her other interests include watercolor painting, photography, music, cooking and gardening.